COAL CONVERSION TECHNOLOGY

Coal Conversion Technology

I. Howard-Smith
G.J. Werner

NOYES DATA CORPORATION

Park Ridge, New Jersey   London, England

1976

Published in the United States of America by
Noyes Data Corporation
Noyes Building, Park Ridge, New Jersey 07656

FOREWORD

The industrialized countries of the world are showing renewed and increasing activity in the area of synthetic fuels production from coal. Existing commercial coal conversion processes, such as the Fischer-Tropsch synthesis, the destructive distillation of coal, methanation etc. resulting in liquid and gaseous fuels, are again becoming economically competitive with petroleum.

The many processes and techniques of coal conversion have as a basic concept the transmutation of coal into forms acceptable to our transportation and heating equipment. To accomplish this, high-sulfur coals must be desulfurized, high-ash coals must be demineralized, and most important of all, solid coal must be depolymerized into liquid and gaseous products that can be ignited and burned with facility.

Over 100 processes, in various stages of development, are available to carry out these procedures and are discussed fully in this book.

Originally prepared as an "in-house" report for Millmerran Coal Pty. Ltd. of Brisbane, Australia, it is intended to provide readily accessible and concise data on all major activities in coal conversion technology. A special feature are the flow charts which clearly illustrate all the principal processes.

PREFACE AND ACKNOWLEDGMENTS

The basis of this book was a desire to prepare a report which provides readily accessible and concise data on all major activities in coal conversion technology, as it had been our experience that no single book was available which was completely up to date and which dealt exclusively with this subject.

It is our hope that this report will be of benefit to those now involved in coal conversion and will also lead to a general increase of interest in this field.

To the best of our knowledge we have compiled information on all the major processes, both past and present, though, in certain instances, various techniques have not been detailed due to either a lack of meaningful data, or if a particular non-commercial process bore a close resemblance to a more well-known, commercial technique, in this case, reference has been made to the "modified" non-commercial processes in the various sections.

To the numerous groups and companies who have provided data, reports and advice; to members of the Queensland University, Department of Chemical Engineering; Officers of the Queensland Department of Commercial and Industrial Development; Queensland Department of Mines; Queensland Coal Board and all others who have given invaluable assistance, we tender our sincere appreciation.

Specific acknowledgement must be made to Mr. W.G. Glidden for the preparation of the charts and his unfailing interest and attention to the completion of this report.

As in any work of this nature the comments expressed are those of the authors and no reflection on the views of any individual, organisation, company or authorative body is intended.

I. Howard-Smith
G.J. Werner
Millmerran Coal Pty. Ltd.
Brisbane, Australia

INTRODUCTION

It is pointless to reiterate the oft presented case for industrialised nations to turn their attention towards the efficient utilisation of coal resources. Both government and industry are already demonstrating increased activity in the area of synthetic fuels production from coal. Existing commercial coal conversion processes, e.g. Fischer-Tropsch synthesis, carbonisation –tar hydrogenation, methanol and ammonia syntheses, are becoming economically competitive with petroleum-based derivatives. The "second generation" processes, e.g. S.N.G. production, solvent extraction-hydrogenation and direct catalytic hydrogenation, are, to a certain extent, already entering the commercial phase, notably the production of substitute natural gas via Lurgi coal gasification with methanation.

Escalating prices for petroleum crude oil and its derivatives will have an important and positive effect on the commercial viability of synthetic fuels from coal, but this consideration should not represent the only variable in the energy equation. Of equal significnace is the value of energy independence to a nation in the strategic sense, to serve as some insulation against the dependence of supply from foreign sources and the vagaries of political and diplomatic affiliations. In short, a premium may be well worthwhile to secure a stable energy base and to restrain the worsening balance of payments resulting from a major dependence on foreign sources.

Coal Conversion Processes

The many processes and techniques of coal conversion have, as a basic concept, the representation of coal in acceptable socio-economic forms. To accomplish this, high-sulphur coals must be desulphurised, high-ash coals must be demineralised, and solid coal must be depolymerised into conventionally-acceptable liquid and gaseous products. Processes, in various stages of development, are available to carry out these procedures and are discussed in the following sections.

1.   Desulphurisation.

Many techniques are available for the desulphurisation of coal, coal tar and impure coal gases. High-sulphur coals have their pyritic sulphur removed by chemical leaching, such as with ferric chloride solution, in a number of proprietary processes, e.g. C.S.I.R.O., and Kennecott's Legemont Laboratories processes. The removal of organic sulphur from coal, coal extract or tar is usually by hydrogenation, in the form of hydrogen sulphide. Sulphurous impurities in gases are removed by water scrubbing (initial treatment) and subsequent purification, as in Lurgi's Rectisol process, which employs a cold methanol solvent, or the Claus Process, which produces dry, elemental sulphur from hydrogen sulphide gas. A hot-desulphurisation process is currently under development by U.S. Steel Corp. which eliminate the need for cooling and wet scrubbing of coal gasifier gas for sulphur removal.

It will probably be more efficient and, in the long run, less expensive, to desulphurise high-sulphur fuels before use rather than resort to stack-gas cleaners to reduce sulphurous pollutants in order to conform with government pollution standards.

2.   Demineralisation and Depolymerisation.

The removal of mineral matter (ash) from coal is best achieved by the process of solvent extraction. The organic material in the coal is taken into solution by either a hydrogen-donor solvent (e.g. tetralin), or a coal-derived (process-derived) solvent in the presence of hydrogen or synthesis gas. The macro-molecules of the coal substance are initially thermally disintegrated, under the temperatures and pressures of reaction, yielding reactive fragments or free radicals which are stabilised by the addition of hydrogen. This hydrogen is derived from either a hydrogen-donor solvent (which itself becomes more aromatic in the process), directly from gaseous hydrogen, or nascent hydrogen produced by a water-gas shift reaction of carbon monoxide present in the synthesis gas. With solvent extraction employing a process-derived solvent (e.g. anthracene oil derivatives), hydrogen must be added to the reactor, or the reactive coal fragments (or free radicals) will re-polymerise to tars and precipitate from the solution. This latter method, employing a process-derived solvent with gaseous hydrogen or synthesis gas, is the basis of the Solvent Refined Coal Process. These hydrogenation reactions may be catalysed by commercially available catalysts such as tungsten sulphide, tungsten-nickel on alumina, and various molybdates.

There is some experimental evidence to suggest that the mineral matter in the coal may catalyse the hydrogenation reaction and this is worthy of further study. Up to 95%, (or even greater), of the fixed carbon in some coals can be solubilised by the solvent extraction-hydrogenation process. The insoluble residue, ash plus unreacted coal, is separated from the coal solution by filters, centrifuges, hydrocyclones or by vapour stripping. The filtrate may then be solidified or further catalytically hydrotreated (hydro-desulphurised and hydro-cracked) to yield low-sulphur liquid products equivalent to petroleum products.

Early work on the solvent extraction of coal was carried out by numerous European researchers, most notably A. Pott and F. Broche in Germany. F. Bergius developed the first commercial process (now obsolete) and much of Germany's war-time production of aviation fuels was derived from Bergius plants. Renewed synthetic fuels interest in the 1960's initiated the more recent processes of the Pittsburgh & Midway Coal Mining Co., Universal Oil Products Co., Consolidation Coal Co., Mitsui & Co., Old Ben Coal Co., British National Coal Board (Coal Research Establishment) and Australia's C.S.I.R.O.

While the process is still in the early stages of development, with numerous pilot plant evaluation studies underway, some companies have already announced plans for the construction of commercial-scale plants based on one or other of the various processes under development. (e.g. Mitsui; Wheelabrator-Frye, Gulf Oil, Southern Services, Inc.)

A great advantage of the solvent extraction-hydrogenation processes is the ability to produce a "tailor-made", clean material suitable for further processing into a wide range of products such as anode carbon, carbon fibre, formed-coke, solid combustion fuel, synthetic crude oil and substitute natural gas. All these products would be extremely low in impurities, as up to 90 percent of the sulphur in the original coal feedstock is removed, either as an insoluble residue, or as hydrogen sulphide gas by hydrogenation.

3.   Carbonization (also called pyrolysis, destructive distillation, or volatilization.)

Thermal decomposition of coal in the absence of air (under vacuum or in an inert atmosphere) results in the evolution of gases and liquids. The gases are composed of mixtures of hydrogen, carbon monoxide, carbon dioxide, methane and higher hydrocarbons, and water. The liquid products are condensed to tars, which represent the products of repolymerisation of free radicals formed during the thermal decomposition of the coal. Should the tars be catalytically treated in the presence of hydrogen, the tar molecules are hydrogenated to simpler aromatic molecules, and this provides an important source of benzene, toluene and xylene.

The relative yields of liquid and gaseous pyrolysis products depend on the proportions of volatile matter present in the coal, the reaction temperature and residence time of the coal particles within the reaction zone. A significant increase in the yield of volatiles can be obtained by the technique of "flash" pyrolysis, which entails rapid heating of the coal particles (of the order of thousands of degrees per second) with consequent very short residence times (of the order of seconds or fractions of a second) to inhibit the further decomposition or polymerisation of the products of pyrolysis. To optimise tar yields, the time spent within the reaction zone must be limited to prevent the thermal decomposition of the tars to gaseous products. Very high temperatures must be employed to produce an optimum gas yield under flash pyrolysis conditions. Various avenues have been investigated to produce gases by rapid, high temperature pyrolysis, including the utilisation of laser irradiation, electric arc and plasma jets. The rapid heating technique minimises the repolymerisation of the pyrolysis products in the coal particles and usually results in the production of acetylene-rich gases.

The devolatilised material which remains is a char of higher calorific value than the original coal, with properties suitable for utilisation in gasification or combustion.

4.   Gasification.

A wide range of commercially-available processes exist for the gasification of coal (or char). Generally, the gasification of coal entails the controlled, partial oxidation of the coal to convert it to a desired product. The coal is heated either directly by combustion, or indirectly by an external heat source such as a D.C. current or recirculated hot char or ash agglomerates. The gasifying media are passed into intimate contact with the heated coal and react with the carbon or with the primary products of thermal decomposition to produce the desired gaseous product.

There are, generally speaking, broad relationships between the gasification reactants and products and these are represented in the following table:

Type of Gasifier	Reactants	Product
Single vessel	Coal, air	Low-BTU, nitrogen-rich producer gas. (100-150 BTU/SCF)
Single Vessel	Coal, Air, Steam.	Semi-water gas (200 BTU/SCF)
Single vessel (two stage, cyclic.)	Coal-Air (Blow phase) Coke-Steam (Make phase)	Water gas (synthesis gas); mixture of $CO + H_2$ (250-350 BTU/SCF)
Single vessel	Coal, Oxygen, Steam	Synthesis gas; mixture of CO, H_2, minor methane (250-500 BTU/SCF)
Hydrogasifier (external H_2 source)	Coal-Hydrogen/Char Hydrogen	Methane-rich gas for S.N.G. production by methanation
Hydrogasifier (Two stage)	Stage 1. Char, Oxygen, Steam Stage 2. Coal, Steam	Synthesis gas (400 BTU/SCF)
Two-vessel (Gasifier-Combustor)	Gasifier: Coal-Steam Combustor: Char-Air	Synthesis gas (400 BTU/SCF) Flue gas
Single vessel (External heat source)	Coal, Steam (optionally catalytic)	Synthesis gas

The low-BTU gas can be utilised as a fuel gas in a power station or industrial plant located adjacent to the gas plant. The low heating value (100-150 BTU/SCF) of this nitrogen-rich gas precludes the economic transportation of the gas via a pipeline over any distance. The gas, free of particulates and corrosive impurities, would also be suitable for powering a gas turbine.

The medium-BTU gas (250-500 BTU/SCF) consists largely of carbon monoxide and hydrogen and as such, is suitable for·the catalytic synthesis of a wide range of products e.g. Fischer-Tropsch synthesis for oil, gasoline, L.P.G. and wax; methanol and ammonia synthesis; Oxo compounds; and the production of substitute natural gas by catalytic methanation. This latter process involves the upgrading of the medium-BTU synthesis gas to a methane-rich gas of pipeline quality,(about 950-1050 BTU/SCF, dehydrated and pressurised to 1,000 p.s.i.) by the reaction of hydrogen and carbon monoxide over heated nickel catalysts to form methane and steam. Before the catalytic methanation, the ratio of hydrogen to carbon monoxide in the gas is changed to 3:1 by a CO-shift conversion reaction.

The production of low- and med-BTU gases i.e. fuel gas and synthesis gas, respectively, is carried out commercially on a large scale. Many plants have been constructed to produce ammonia and methanol from coal via coal gasification-gas synthesis. A large plant at Sasolburg in South Africa operates economically to produce a wide range of liquid and gaseous products by Fischer-Tropsch synthesis and a second, larger plant, is currently on the drawing board. (see page 16).

The production of substitute natural gas has been successfully demonstrated on a small commercial scale at Westfield, in Scotland, utilising a Lurgi "Hot Gas Recycle" methanator to upgrade the heating value of synthesis gas,from Lurgi steam-oxygen pressure gasifiers,to pipeline quality for distribution through an existing commercial pipeline system.

In the United States, coal gasification projects, predominantly based on Lurgi commercial technology, are planned for the production of substitute natural gas from coal. The first plant, to produce 250 million S.C.F./day of 975 BTU/SCF gas,is expected to be operational in 1977. Although it is anticipated that the cost of the coal-derived pipeline gas will be higher than that for currently- available natural gas, the shortfall in the supply of domestic natural gas,in the United States,will stimulate the consumers to pay the extra cost of service involved in the production of substitute natural gas from coal.

In conclusion, it should be stated that it is expected synthetic fuels from coal will become increasingly important as an indigenous source of energy to many nations having significant reserves of coal. The rate at which these synthetic fuels will increase in importance will depend to a large extent on well-coordinated research and development efforts by industry, under the auspices of governmental authorities, and the efficient exchange of information between the various organisations engaged in coal conversion technology on somewhat more than an "ad hoc" basis.

ACRONYMS

ACIRL	Australian Coal Industries Research Laboratories
AEC	Atomic Energy Commission
AGA	American Gas Association
AMAX	American Metals Climax
AMOCO	American Oil Company
ATC	Applied Technology Corporation
BCR	Bituminous Coal Research, Inc.
BCURA	British Coal Utilisation Research Association
BGC	British Gas Council, The
BMI	Battelle Memorial Institute
B & W	Babcock and Wilcox Ltd.
C — E	Combustion — Engineering, Inc.
CEGB	Central Electricity Generating Board
CHEMICO	Chemical Construction Corporation
CONOCO	Continental Oil Company
CONSOL	Consolidation Coal Company
CRE	Coal Research Establishment
CSIRO	Commonwealth Scientific and Industrial Research Organisation
EPA	Environmental Protection Agency
EPRI	Electric Power Research Institute
ERDA	Energy Research and Development Administration
ERI	Eyring Research Institute
F — T	Fischer — Tropsch
F — W	Foster — Wheeler Corp.
GDR	German Democratic Republic
GEMCO	Gulf Energy and Minerals Company
GRD	Garrett Research and Development Company
HRI	Hydrocarbon Research, Inc.
ICI	Imperial Chemical Industries Ltd.
IFE	International Furnace Equipment Company Ltd.
IGI	Il Gaz Integrale
IGT	Institute of Gas Technology
IOL	International Oil Limited
ISC	International Systems and Controls Corp.
K — T	Koppers — Totzek
MITI	Ministry of International Trade and Industry (Japan)
NCA	National Coal Association
NCB	National Coal Board
NCRAC	National Coal Research Advisory Committee
NRDC	National Research and Development Committee
NSF	National Science Foundation
OCR	Office of Coal Research
PAMCO	Pittsburgh and Midway Coal Mining Company
QIT	Queensland Institute of Technology
SASOL	South African Coal, Oil and Gas Corporation
SOHIO	Standard Oil of Ohio
SRI	Stanford Research Institute
TOSCO	The Oil Shale Corporation
TVA	Tennessee Valley Authority
UOP	Universal Oil Products Company
USBM	U.S. Bureau of Mines

GLOSSARY OF TERMS

Acceptors: Calcined carbonates which absorb carbon dioxide evolved during gasification, exothermically.

Acid Gas Removal: A section of a gas plant where hydrogen sulphide and carbon dioxide are removed from the gas stream.

Activated Carbon: Carbon obtained by carbonization in the absence of air, preferably in a vacuum; has the property of absorbing large quantities of gases, solvent vapours; used also for clarifying liquids.

Agglomerate: Assemblage of ash particles rigidly joined together, as by partial fusion (sintering).

Anode Carbon Carbon of high purity, usually crystallised to graphite form, widely used in Leclanche cells, in rods for alumina refining, in electric arcs and nuclear reactors.

Autoclave: A vessel, constructed of thick-walled steel (alloy steel or nickel alloys), for carrying out chemical reactions under pressure and at high temperatures.

Bench Scale Unit: A small-scale laboratory unit for the testing of process concepts and operating criteria as a first step in the evaluation of a process.

Binder: Carbon products, tars etc. used to impart cohesion to the body to be formed; a coal-extract binder may be used to prepare formed-coke pellets from non-coking coals.

Btu: Abbreviation for British Thermal Unit (= 1.05506 kJ.)

B.T.X. Benzene, toluene, xylene; aromatic hydrocarbons; toluene is methyl-benzene , xylene is dimethyl benzene.

Bunker Oil: Heavy fuel oil for ships.

Caking coal: Coal which cakes, or agglomerates, or forms coke, when heated.

Calcine: Carbonate which has been heated to drive off carbon dioxide.

Carbon Fibre: Very fine filaments of the order of 8 microns diameter which are used in composite materials, being bound with resins.

Carbonization: The destructive distillation of coal in the absence of air accompanied by the formation of char (coke), liquid (tar) and gaseous products.

Catalyst: A substance which accelerates the rate of a chemical reaction without itself undergoing a permanent chemical change.

Centrifuge: An apparatus rotating at high speed to separate solids from liquids, e.g. undissolved residue from coal solution in the S.R.C. process.

Char: The solid residue from the carbonization of coal after the volatile matter of the coal has been driven off by heating.

Claus Process: A process for recovering elemental sulphur from hydrogen sulphide gas, utilising a brick-work kiln, at high temperatures with oxygen reacting with the H_2S to yield dry sulphur and steam.

Combustion Gas: Gas formed by the rapid oxidization of coal. e.g.; burning.

Combustor: A vessel in which combustion of gaseous products from a fuel takes place by the chemical union of oxygen with the gas.

CO Shift Conversion: See Shift Conversion

Crude Gas: Gas produced in a gasifier containing a wide range of impurities, also known as off-gas.

Cyclone: A conical vessel used to classify dry powders or extract dust, or separate dust from a gas stream, by centrifugal action.

D.A.F.:	Dry, ash-free (abbrev. for); a term which relates to the organic fraction in coal; the U.S. terms "moisture, and ash-free (MAF)" and "moisture, and mineral matter free (m,mmf) are equivalents.
Degasification:	A process for draining naturally-occurring methane from coal seams.
Delayed Coking:	A process whereby coal is subjected to a long period of carbonization at moderate temperatures to form metallurgical coke.
Demineralization:	Removal of mineral matter (ash) from coal by solvent extraction, usually under hydrogen atmosphere.
Demonstration Plant:	A plant whose design is based on data derived from pilot-scale testing, of sufficient capacity to demonstrate the large-scale feasibility of a process.
Depolymerization:	The change of a large molecule (e.g. coal polymers) into simpler molecules (e.g. aromatics, B.T.X.), usually accompanied by the substitution of hydrogen for oxygen.
Destructive Distillation:	The distillation of coal or other solids accompanied by their decomposition; destructive distillation of coal yields coke, tar, ammonia, gas etc.
Desulphurization:	The removal of sulphur from hydrocarbonaceous substances by chemical reactions. Various processes are Claus, Appleby-Frodingham, C.S.I.R.O., ferric chloride leaching, Kennecott.
Devolatilization:	The removal of a proportion of the volatile matter from medium-and high-volatile coals to prevent subsequent caking.
Dissolution:	The taking up of a substance by a liquid with the formation of a homogeneous solution.
Distillation:	A process of evaporation and re-condensation used for separating liquids into various fractions according to their boiling points or boiling ranges.
Ebullated Bed:	A boiling bed; gas, containing a relatively small proportion of suspended solids, bubbles through a higher density fluidised phase with the result that the system resembles a boiling liquid.
Effluent Gas:	Gas issuing from a gasifier or combustor.
Electrode Carbon:	See Anode Carbon.
Entrained Bed:	A bed in which solid particles are suspended in a moving fluid, and are progressively carried over in the effluent stream.
Entrained Flow:	See Entrained Bed.
Extraction:	A process for dissolving certain constituents of a mixture by means of a liquid with solvent properties for selected components only.
Extraction Gas:	See Gas Extraction in Section I of this report.
Extraction-Hydrogenation:	Extraction carried out in the presence of hydrogen either as a gas or derived by transfer from H-donor solvents.
Extractive Coking:	Similar to Delayed Coking process, with the emphasis on high tar yields to produce liquids.
Filter Cake:	The moist residue remaining from the filtration of a slurry to produce a clean filtrate.
Filtrate:	A liquid free of solid matter after having passed through a filter.
Filtration:	The separation of solids from liquids by passing the mixture through a suitable medium, e.g. cloth, paper, diatomite.
Fischer Assay:	An assay for the determination of oil (tar) yields from coal or oil shale; conducted in a retort under an inert atmosphere with a gradual increase in temperature.
Fischer-Tropsch Catalyst:	Iron and cobalt catalysts developed by Fischer and Tropsch for the catalytic synthesis of liquid fuels from coal-derived synthesis gas.

Fixed Bed:	A bed through which gases pass at a velocity low enough so as not to blow the solid particles from the bed.
Flash Carbonization:	A carbonization process characterised by very short residence times of coal in the reactor to optimise tar yields; also called flash pyrolysis.
Fluidized Bed:	A bed through which a fluid is passed with a velocity high enough for the solid particles to separate and become freely supported in the fluid.
Flue Gas:	Gas issuing from a combustor; either exhausted to atmosphere or expanded through a gas-turbine.
Fly Ash:	A fine ash from the pulverized coal burned in power station boilers, or entrained ash carried over from a gasifier.
Fractionation:	Distillation process for the separation of the various components of liquid mixtures; an effective separation can only be achieved by the use of fractionating columns attached to the still; also called Fractional Distillation.
Freidel Crafts Catalyst:	The catalyst employed in the synthesis of benzene hydrocarbons by the action of alkyl halides on aromatics in the presence of anhydrous aluminium chloride.
Fuel Cell:	A galvanic cell in which the oxidation of a fuel (e.g. coal) is utilised to produce electricity.
Fuel Gas:	Low heating value product generally utilized on-site for power generation or industrial use.
Gasification:	Conversion of solid or liquid hydrocarbon substances to a gaseous phase in a gasifier.
Gasifier:	A vessel in which gasification occurs usually utilizing fixed bed, fluidized bed or entrained bed units.
Gas Oil:	A petroleum distillate obtained after kerosene and before lubricating oil; also called diesel oil.
High-BTU Gas:	Methane-rich gas product with an H.H.V. between 900-1,000 BTU/SCF (See Substitute Natural Gas).
H.H.V.:	Higher heating value; calorific value of a unit volume of a gas.
Hydro coking:	Coking of tars, S.R.C., etc. under hydrogenating conditions to form liquid products.
Hydrocracking:	A thermal process in which petroleum distillates or coal-derived extracts are broken down into products of lower boiling range with the addition of hydrogen by a catalytic reaction.
Hydrocyclone:	A small cyclone extractor for removal of suspended solids from a flowing liquid by means of the centrifugal forces set up when the liquid is made to flow through a tight conical vortex.
Hydrogasification:	Gasification that involves the addition of hydrogen to the products of primary gasification to optimise formation of methane.
Hydrogenation:	Chemical reactions involving the addition of hydrogen, present as a gas, to a substance in the presence of a catalyst under high temperatures and pressures.
Hydrogen-Donor Solvent:	Solvent such as anthracene oil, tetralin (tetrahydronaphthalene), decalin, etc. which transfer hydrogen to coal constituents causing depolymerisation and consequent evolution of liquid products of lower boiling range which are then taken up by the solvent.
Industrial Gas:	See Fuel Gas.
Inerts:	Macerals in coal not readily changed by the action of solvents in the solvent extraction of coal, e.g. fusinite.
In-Situ:	In its original place, e.g. underground gasification of a coal seam.
Intermediate BTU Gas:	Synthesis gas product with a H.H.V. between 250-500 BTU/SCF consisting mainly of carbon monoxide and hydrogen.

Liquefaction:	Conversion of a solid to a liquid; with coal this invariably involves hydrogenation to depolymerise the coal molecules to simpler molecules.
Low BTU Gas:	Nitrogen-rich gas product with an H.H.V. between 100-200 BTU/SCF produced in air-blown gasifiers and air injected "in-situ" gasification. Also known as "Producer Gas".
L.P.G.:	Liquefied petroleum gases such as propane and butane; used for fuels.
Medium BTU Gas:	See Intermediate BTU Gas.
Methanation:	Process for the production of methane by passing carbon monoxide — hydrogen mixtures over nickel catalysts; according to the reaction, $CO + 3H_2 \rightarrow CH_4 + H_2O$.
M.H.D.:	Magnetohydrodynamic generator, (abbrev. for); device which produces electrical energy from an electrically-conducting gas (plasma) flowing through a transverse magnetic field; also called magnetoplasmadynamic generator (M.P.D. generator).
Natural Gas:	Naturally occurring gas extracted from sedimentary structures consisting mainly of methane and having an H.H.V. to 1,050 BTU/SCF.
Oxo:	The radical $O =$ in organic compounds, e.g. ketones.
Petrochemicals:	Those derived from crude oil or natural gas, or their coal-derived substitutes; they include light hydrocarbons such as butylene, ethylene and propylene, the raw materials for the production of plastics by polymerisation.
Phenols:	A group of aromatic compounds having the hydroxyl group directly attached to the benzene ring. They give the reactions of alcohols, forming esters, ethers and thiocompounds; phenols are more reactive than the benzene hydrocarbons; derived from coal tar.
Phenosolvan Process:	Process for the dephonolisation of phenolic effluents from coke-oven, carbonisation, gasification, hydrogenation and synthetic phenol and resin producing plants. The phenolic effluents are treated with an oxygen-containing organic solvent (Phenisol or Phenosolvan) in multi-stage counter-current extractors whereby the phenols are absorbed by the solvent. The extract is separated into crude phenols (phenol, cresols, xylenols, higher phenols) and solvent by distillation. Developed by Lurgi GmbH.
Pilot Plant:	Plant of a scale intermediate between laboratory-scale and large-scale units to evaluate integrated, continuous-operation parameters in the development of a new process, or to provide engineering design data for a commercial-scale plant.
Pipeline Gas:	A methane-rich gas that conforms to certain standards and having a H.H.V. between 950-1,050 BTU/SCF. Standards include minimum water content, minimum inert gases, minimum H_2 and CO content and compressed to 1,000 p.s.i.g.
Pre-coat:	Layer of suitable filtering medium, e.g. diatomaceous earth, laid down on a rotary filter cloth prior to operation.
Process Development Unit:	Small-scale plant operated to provide information on the operation of process units, to provide design data for larger-scale plants; similar to pilot plant.
Purification:	Removal of the wide range of impurities present in gases from coal gasification to yield purity gas. See Recitsol Process.
Pyrolysis:	The decomposition of coal by heat. See Carbonization.
Quenching:	Cooling by immersion in oil or water bath or spray.
Raney Nickel Catalyst:	Nickel sponge used as a catalyst in the hydrogenation of organic materials and the methanation of synthesis gas to methane.
Raw Gas:	See Crude Gas.
Reactivity:	Susceptibility to chemical change; in coal conversion, the reactivity of the coal for conversion to liquid products is a function of the D.A.F. volatile matter content and the petrographic composition of the coal.

Reactor:	Vessel in which coal-conversion reactions take place.
Rectisol Process:	A process, developed by Lurgi GmbH, for the purification of coal-gasification gas based on the capability of cold methanol to absorb all gas impurities in a single step; gas naphtha, unsaturated hydrocarbons, sulphur, hydrogen sulphide and carbon dioxide are removed from the gas stream by the methanol at temperatures below 0^oC.
Reducing Gas:	Used as a reducing agent in redox reactions, e.g. hydrogen, super-heated steam.
Reforming Processes:	A group of proprietary processes in which low-grade or low molecular weight hydrocarbons are catalytically reformed to higher-grade or higher molecular weight materials; also applies to the endothermic reforming of methane, for the production of hydrogen by the reaction of methane and steam in the presence of nickel catalysts.
Residence Time:	Time spent by a typical particle in a reaction zone.
Semi-Water Gas:	A mixture of carbon monoxide, carbon dioxide, hydrogen and nitrogen obtained by passing an air-stream mixture through an incandescent bed of coke; H.H.V. about 120 BTU/SCF.
S.C.F.:	Standard cubic foot (abbrev. for); ($=0.0283168 m^3$).
Scrubber:	Apparatus in which a gas stream is freed of tar, ammonia and hydrogen sulphide.
Shift Conversion:	Process for the production of gas with a desired carbon monoxide content from crude gases derived from coal gasification; a CO-rich gas is saturated with steam and passed through a catalytic reactor where the CO reacts with steam to produce hydrogen and carbon dioxide, the latter being subsequently removed in a wash plant. The ratio of hydrogen to carbon monoxide in the product gas can be changed at will.
Slagging:	Operating above the ash-fusion temperature so that the ash is removed from the gasifier as a molten slag.
Slurry:	A thin paste produced by mixing finely divided solids in a liquid medium; sufficiently fluid to flow viscously.
Solvation:	The association or combination of molecules of solvent with solute ions or molecules.
Solvent:	That component of a solution which is present in excess, or whose physical state is the same as that of the solution.
Solvent Extraction:	Selective transfer of desired coal constituents from finely divided coal particles into a suitable solvent after intimate mixing, usually at high temperatures and pressures in the presence of hydrogen, with or without a catalyst, followed by phase separation.
Solvent Refined Coal:	A coal extract derived by solvent extraction; a brittle, vitreous solid (M.P. 300°-400^oF) containing about 0.1% ash and about 10% of the sulphur in the original coal feedstock; calorific value is about 16,000 Btu/lb. May be used as a clean fuel for power generation by combustion, utilised for the production of high-grade metallurgical coke, anode carbon and activated carbon by coking, or hydrogenated to produce synthetic crude oil. (abbrev. S.R.C.).
Substitute Natural Gas:	A synthetic gas conforming to natural gas standards. (abbrev. S.N.G.)
Syncrude:	Synthetic crude oil; oil, produced by the hydrogenation of coal or coal extracts, which is similar to petroleum crude.
Synthesis Gas:	Gas consisting of mixtures of carbon monoxide and hydrogen in proportions that allow for synthesis to higher molecular weight products, e.g. methane and Fischer Tropsch products.

Tail Gas: A gas issuing from a gas-treatment unit which may be recycled to the process or exhausted.

Tar: The products of distillation or high—or low— temperature carbonization of coal; coal tar consists of hydrocarbon oils (B.T.X. and higher homologues), phenols, and bases such as pyridine, quinoline pyrrole and their derivatives; gas tar is tar condensed from coal gas and provides, by distillation, ammoniacal liquor, benzole, naphtha and creosote with a residue of pitch which is dehydrated to produce bitumen.

Therm: A unit of heat used as a basis for the sale of town and natural gas; equal to 100,000 Btu.

Town gas: Gas product with an H.H.V. of 600 BTU/SCF and conforming to certain standards including prescribed water content, inert gases, H_2, CO, methane content and Wobbe number.

Water Gas: Gas produced by the reaction of carbon and steam to provide carbon monoxide—hydrogen mixtures;similar to synthesis gas.

Working Fluid: A gas stream which directly does work, e.g. powering a gas turbine.

TABLE OF CONTENTS

Section I

COAL CONVERSION PROCESSES FOR THE PRODUCTION OF SOLID, LIQUID AND GASEOUS PRODUCTS

Section II

HIGH BTU COAL GASIFICATION PROCESSES

Section III

LOW AND MEDIUM BTU
COAL GASIFICATION PROCESSES

SECTION I

Coal Conversion Processes
for the production of
Solid, Liquid and Gaseous Products

BERGIUS

Process Developer:	F. Bergius.
Description:	Coal is first ground to a fine size and mixed with a process-derived hydrocarbon liquid and a catalyst. This mixture is reacted with hydrogen (produced by the gasification of coal) at pressures up to 10,000 p.s.i. The products from the first reactor are separated into light, middle and bottom fractions.

The middle fraction is further treated over a catalyst in a vapour phase and under relatively mild conditions to produce petroleum–like products. The bottom fraction is filtered to remove solids (unreacted coal, catalyst and ash) and the remaining liquid utilised as a mix with fresh coal being processed in the first reaction.

Operating Conditions:

Process Type	Temp. °F.	Press. p.s.i.	Vehicle	Catalyst	Product
Solution-hydrogenation	900	3,000-10,000	Recycle Oil	Iron Oxide	Light Oils

Status:
The liquefaction of coal via hydrogenation at high temperatures and pressures was intensively studied by F. Bergius over the period 1910 to 1927. After World War I, the I.G. Farben syndicate began work independantly and by 1927 German development had passed entirely into their hands. In that year, the first commercial plant was started in Leuna and by 1939 that plant had reached a production in the vicinity of 400,000 tons of oil a year. By September, 1939, seven hydrogenation plants had been erected in Germany, with a total capacity of 1.4 million tons of oil and liquefied gas a year.

During World War II, several additional plants were constructed until by 1944 there were 18 hydrogenation plants in operation with a capacity for production of four million tons a year of oil. In fact, practically the entire production of aviation fuel in 1944 was from these hydrogenation plants.

Research and developmental work on the process was also carried out by Japanese interests (i.e. Mitsui at Fusan, Manchuria 1928) and the British Fuel Research Station, London, in the 1930's. In Billingham, U.K., a plant for the hydrogenation of coal and tar had an annual production of 150,000 tons of liquid fuel. Another plant in Korea was operational in 1942 to produce 110,000 tons of gasoline and diesel fuel annually.

Immediately after World War II, there was an oil shortage in the U.S.A. The Office of Synthetic Liquid Fuels was established in 1944 and expended U.S. $87.6 million in the following eleven years, investigating the production of liquid fuels from a number of resources from coal to corn waste. Coal research liquefaction plants were built in what are now the Pittsburgh and Bruceton Energy Stations. Additional research was done at the Morgantown and Grand Forks Stations. At Louisiana, Mo., a large government ammonia plant was converted into two liquefaction demonstration plants one of which utilised the Bergius process to produce about 200 barrels per day of gasoline.

No commercial Bergius plants are currently operating.

CATALYTIC COAL LIQUEFACTION (C.C.L.).

Process Developer:	Gulf Research and Development Co., a subsidiary of Gulf Oil Corp.
Technical Services:	Blaw-Knox Chemical Plants Division, Dravo Corp.

| Description: | Coal, dried and pulverised, is mixed with a process—derived oil and the resultant slurry is forced up through an arrangement of fixed catalysts by a stream of hydrogen at about 2,000 p.s.i. pressure. Liquid product may be utilised for the production of gasoline or diesel fuels, or petrochemical feedstock. |

Operating Conditions:

Reactor	Temp. °F	Press. p.s.i.	Reactants	Products
Catalytic, fixed bed	900	2,000	Coal-oil-H_2	Synthetic Crude Oil and Gas

Status:

Gulf Research and Development Co. have been testing this process for eight years on an experimental bench-scale level. In January, 1975 a one ton per day pilot plant began operation. This U.S. $1.5 million plant has been designed to produce a yield of 3 barrels of oil from each ton of D.A.F. coal. Operation of the pilot plant is to provide design data for a larger demonstration plant, a conceptual design of which is currently being prepared. The pilot plant is located at Harmarville.

Under the sponsorship of the RANN Division of the National Science Foundation, Gulf R & D and the College of Earth and Mineral Sciences, The Pennsylvania State University, are jointly undertaking studies on the dependence of coal liquefaction behaviour on coal characteristics. Various bench-scale units are operated under this study to investigate catalytic coal liquefaction. Results to date have indicated that the highest conversion yields are with high-ash, high-volatile bituminous coals containing high proportions of the reactive macerals, vitrinite and exinite.

Note: Gulf Oil Corporation is also conducting R & D on the S.R.C. process through its subsidiary, the Pittsburgh and Midway Coal Mining Company, now combined, along with Gulf Mineral Resources into a synthetic fuels research division — GEMCO (Gulf Energy and Minerals Co.)

CLEAN—COKE PROCESS

Process Developer: U.S. Steel Corp., Research Laboratory, Monroeville, Pa.

Sponsor: Office of Coal Research.

Description:

Coal, after beneficiation and sizing in a coal-preparation plant, is split into two fractions. Portion of the coal is processed through a carbonisation unit where it is devolatilised and partially desulphurised to produce the char that serves as the base material for the production of metallurgical coke. The remainder of the coal feed is slurried with a process-derived carrier oil and is hydrogenated to convert a large proportion of the coal to liquid.

Liquid products from both carbonisation and hydrogenation are composited and processed through a central liquids—treatment unit. In this unit, the liquids are converted into low-sulphur liquid fuels, chemical feedstocks, and three oil fractions that are recycled to other areas of the process.

One of the recycle fractions is used primarily as a carrier oil for the hydrogenation reaction. A second recycle oil is sent to the carboniser where it is converted to pitch coke. The pitch coke and char mixture is blended with the third recycle oil that serves as a binder, and this mixture is formed into pellets in the coke-preparation unit. These pellets are subsequently baked to produce a formed metallurgical coke with strength properties equivalent to those of blast-furnace coke made by conventional coking.

The coke-preparation cycle, from char production to final coke, is carried out in a closed system with the off-vapours collected and returned to the process. Thus, there are no significant emmissions of volatile matter durmg these operations, and atmospheric pollution is practically non-existent. Gaseous products from all operations are processed through a common system to provide chemical feedstocks, low-sulphur gaseous fuels, and hydrogen for recycle to hydrogenation and liquids treatment.

Operating Conditions:

Process Unit	Temp. °F	Press.p.s.i.	Products
Carbonisation	1,200-1,400	100	Liquid Products, Gas, Char for Coker
Extraction-hydrogenation	900	3,000-4,000	Liquid Products, Gas, Filter Cake for Gasifier

Status:

The Office of Coal Research awarded a 3-year contract on 10th March 1972 to U.S. Steel Engineers and Consultants, Inc., a subsidiary of U.S. Steel Corporation with a research laboratory at Monroeville, Pa. Under the contract, the U.S. Steel Corp. contributed U.S. $1,925,000 and O.C.R. U.S. $4,636,000 to the cost of the project.

Work to be completed under the contract for the Clean-Coke process included:

1. Design, construction and operation of process-development units (P.D.U.'s) capable of processing 500 lb. per day of coal in integrated units.

2. Continuation of bench-scale studies to augment design and operation of the P.D.U.'s.

3. Preparation of the design for a 100 T.P.D. pilot plant (original specifications called for a pilot plant of up to 240 T.P.D.). A principal objective for the second year of this project was completion of the designs and construction of the P.D.U.'s, which were to be operated during the third year of the contract. Information from the P.D.U.'s will provide the basis for the design of the 100 T.P.D. pilot plant.

Detailed designs were completed on five P.D.U.'s:

1. Coal and coke preparation.
2. Carbonisation.
3. Hydrogenation.
4. Slurry-oil preparation.
5. Binder preparation.

Of these, all but the high-pressure section of the hydrogenation P.D.U. had been purchased and installed in early 1974, with the high-pressure hydrogenation equipment completed later in 1974. Startup and evaluation tests are continuing with sufficient experimental work completed for the preliminary design of the pilot plant.

Estimates on the yield of products from a commercial plant processing 17,000 T.P.D. of washed and sized coal (2–3% sulphur) are as follows:

Coal Feed: 5.8 million tons per year.

Annual Production:

Coke	2.2 million tons
Hydrogenation Residue	669,000 tons

Chemicals

Ammonia	34,000 tons
Sulphur	49,000 tons
Ethylene	723 million lb.
Propylene	119 million lb.
Phenol	152 million lb.
o-Cresol	38.4 million lb.
m–,p–Cresol	142 million lb.

Xylenols	150 million lb.
Pyridine	14.6 million lb.
$\propto$ -Picoline	6.1 million lb.
Aniline	29.2 million lb.
Benzene	80.4 million gallons (U.S.)
Naphthalene	229 million lb.

Tar Products.

Creosote Blend Stock 4.13 million gallons (U.S.)
Carbon Black Feedstock 3.7 million gallons (U.S.)

Process loss and fuels consumed account for 29% of the weight of the coal feed.

The basis of this material balance is the processing of the coal feed to produce a maximum of coke pellets and chemicals, with minimum fuels.

The estimate of the time required for construction of a 17,000 T.P.D. commercial plant, following successful completion of integrated pilot plant operations, is three years.

COALCON HYDROCARBONISATION

Process Developer: Coalcon Co. — a joint-venture company of Union Carbide Corp., and Chemical Construction Corp. (Chemico). Chemico is a subsidiary of the General Tire and Rubber Co.'s. Aerojet—General Corp.

Sponsors: Office of Coal Research (50%) and a consortium being formed by Coalcon with several major U.S. companies and state government agencies. Coalcon has letters of intent from the following to join a consortium being formed by Coalcon to concentrate specifically on the needs of the program:

Ashland Oil; Du Pont; Reynolds Metals; Sun Oil, Y & O Coal Co.; Ohio Air-Quality Development Authority; Wood County, West Virginia Development Authority; Belmont County, Ohio; the governors of West Virginia and Ohio.

Other firms are expected to submit letters of intent in the near future.

Description: Sized, dried and preheated coal is fed to a dry, fluidised-bed hydrogenation reactor. Hot vapour products from the reactor go to a gas-liquid product separator where water, liquid fuel product and a complex mixture of gas (H_2, CH_4, CO, C_2-C_5, H_2S, CO_2) are separated to separate streams. The water is sent to a separate AAG unit hydrogen-source gasifier where it reacts with char derived from the hydrogenerator. The AAG hydrogen plant has a gasifier-combustor coupled with char and air in the combustor providing heat for the char-steam reaction in the gasifier. The AAG unit is based on the "Agglomerating-Ash" gasification process. The liquid fuel stream is sent for refining, if desired, with BTX, Phenol, Cresylic Acids and Naphthalene by-products being derived in the process.

The gaseous stream passes to acid-gas removal (with sulphur recovery) then to hydrogen recovery for recycling to the hydrogenerator. Finally, the gas is sent to a hydrocarbon separation unit which delivers separate streams of methane and C_2 - C_5 gases. Lignite, sub-bituminous and bituminous coals can be handled. No details of operating conditions are available.

Status: This process is the result of 30 years of research by Union Carbide in coal liquefaction and hydrogenation. The process has been demonstrated in a 300 T.P.D. semi-works plant and a 25 T.P.D. pilot plant.

In January, 1975 the largest contract ever awarded by the U.S. Interior Department was received by Coalcon Co. This U.S. $237 million contract calls for Coalcon to design, construct and operate a 2,600 T.P.D. demonstration plant using the Hydrocarbonisation Process. The project is to be conducted in four distinct phases over an eight-year period, including operation of the plant for three and a half years.

The program phases are:

Phase I — Conceptual design of the demonstration plant costing U.S. $4 million. Work is already underway.

Phase II — Detailed design valued at U.S. $17 million. (Both phases I and II will be funded solely by the Office of Coal Research.)

Phase III — Plant construction scheduled for completion in 1979.

Phase IV — Operation of the plant for three and a half years.

(**Phases III** and **IV** will cost over U.S. $200 million, to be funded equally by O.C.R. and industry.)

The 2,600 T.P.D. plant will produce 3,900 barrels per day of liquid product and 22 million S.C.F. per day of pipeline gas. The demonstration plant's operation will provide the technology for commercial plants processing 15,000 T.P.D. of coal in the 1980's. Liquid products alone from such a plant would provide the fuel for a 1,000 MW power station. The plant will be located in either Ohio or West Virginia.

COALPLEX

Process Developer: City College of New York, Dept. of Chemical Engineering.

Sponsor: National Science Foundation.

Description: The process is designed to convert coal, in a single operation, to substitute natural gas, lead-free high octane gasoline and low-sulphur coke. Little information is available on the process details.

Status: The concept was originally developed by Arthur M. Squires, chairman of City College's Department of Chemical Engineering. The basic concept envisages a coal-conversion complex producing multiple products at a large saving over the costs of separate production of synthesis gas, S.N.G., gasoline, oil, coke or electricity.

The National Science Foundation has awarded a U.S. $644,400 grant to City College to evaluate the practicability of the concept.

(See Clean-Coke Process, C.O.G. and COGAS.)

COED

Process Developer: FMC Corporation, Research and Development Centre, Princeton, New Jersey.

Sponsors: Office of Coal Research, U.S. Dept. of the Interior; Electric Power Research Institute, Palo Alto, Calif.

Contractors:

1. Blaw-Knox Chemical Plants Division of Dravo Corp. — for project evaluation, estimates, design, procurement and construction of the pilot plant units.

2. Dow Chemical Co. — for evaluation of the COED process as a source of feedstock for a commercial chemicals-complex plant.

3. Chem Systems Inc. — for determination of the value of COED syncrude to new and existing petroleum refineries.

Description:

Coal is crushed, dried and then heated to successively higher temperatures in a series of fluidised-bed reactors. In each fluidised bed, a fraction of the volatile matter of the coal is released. The temperature of each bed is selected to be just short of the maximum temperature to which the coal can be heated without agglomerating and defluidising the bed. Typically, four stages operating at 600°, 850°, 1000° and 1500°F are involved.

The number of stages and the operating temperatures vary with the agglomerating properties of the coal. Heat for the process is generated by burning char in the fourth stage and then using hot gases and the hot char from the fourth stage to heat the other vessels.

The volatile matter released from the coal in the pyrolysis reactors is condensed in a product recovery system. The condensation of the coal-tar vapours is accomplished by direct contact with a water-rich stream in venturi scrubbers. The condensed oil-water mixture is then sent to a gravity decanter vessel for phase separation. Steam-heated dehydrators are used to effect water removal from the pyrolysis oil streams from the decanter. The pyrolysis oil product is filtered by a pressurised, rotary-drum, precoat filter to remove solids representing char fines, which are carried through the cyclones of the fluidised-bed reactors.

The filtered oil product is then mixed with hydrogen and preheated to $700 - 800^\circ$F in the hydrotreating plant. Hydrotreating occurs in a fixed-bed catalytic reactor containing commercial nickel-molybdenum catalyst. Hydrotreating reactions remove sulphur, nitrogen and oxygen from the oil and can produce a $25 - 30^\circ$ API synthetic crude oil product.

Operating Conditions:

PYROLYSIS SECTION	Temp. °F	Press. p.s.i.	PRODUCTS
Fluidised Bed No. 1	600	5 – 10	
Fluidised Bed No. 2	850	5 – 10	Char (59.5%), Oil (19.3%),
Fluidised Bed No. 3	1,000	5 – 10	Gas (15.1%), Liquor (6.1%) *
Fluidised Bed No. 4	1,500	5 – 10	

HYDROTREATMENT SECTION	Temp. °F	Press. p.s.i.	PRODUCT
Catalytic (Ni-Mo) Fixed Bed	750	2,500-3,100	Syncrude. API Gravity 27.2*

* Yields based on Illinois No. 6 — Seam coal. Proximate analysis (dry weight) of this coal is: Inherent Moisture 9.1%; Volatile Matter 37.6%; Fixed Carbon 49.3%; Ash 13.1%.

Status:

The COED process has been under development by FMC Corporation since May, 1962, under the sponsorship of the Office of Coal Research. Successful operation of a 100 pound per hour process development unit led to the design, construction and operation of a pilot plant at Princeton, New Jersey. The pilot plant can process 36 tons of coal per day and hydrotreat 30 barrels of coal-derived oil per day. Startup of the pilot plant began in 1970 and operability of the process has been demonstrated in long-duration runs of over 30 days. Design capacities have been achieved on all parts of the pilot plant with significant accomplishments being demonstrated in solids circulation between multiple fluidised-bed reactors, in the filtration of the product oil, and in upgrading the coal oil to synthetic crude oil through fixed-bed hydrotreating.

6

The COED pilot plant had processed over 18,000 tons of coal to June, 1974. Six different coal types have been tested in the pilot plant, including lignite, sub-bituminous, and high volatile B- and C-bituminous coals. Oil yields ranged from 0.4 to 1.5 barrels per ton of coal, depending on the feed coal type.

Present work will apply added technology to convert the char to a clean fuel gas. The offgas yield is 8,000 to 10,000 S.C.F. per ton of coal. This gas has a H.H.V. of about 500 BTU/SCF. The commercial viability of the COED process relies heavily on saleability of the offgas.

The filtration step has been successfully demonstrated, reducing solids content from about 8% in the raw oil to less than 0.1% in the filtrate.

In the hydrotreatment step, synthetic crude oil has been produced from filtered oils derived from high volatile B-bituminous coals. This synthetic crude oil was evaluated under sub-contracts with Chem Systems, Inc., assisted by Atlantic Richfield and American Oil Company. COED syncrude also was successfully used as a source of bunkering fuel for a U.S. Navy destroyer in 1974 in tests conducted as part of the U.S. Dept. of the Navy's "Project Seacoal". All tests conducted indicated that the COED syncrude can be blended with existing petroleum oils and used as a feed to a refinery. In addition, the COED syncrude can be distilled and sold in two fractions: a naphtha fraction and a low-sulphur No.4 fuel oil fraction. The dollar value of COED syncrude is comparable to that of petroleum crude of similar quality.

The Ralph M. Parsons Company, under separate contract to O.C.R., has started a commercial design for a COED process complex to produce a low-sulphur fuel oil and a clean fuel gas for power generation. This 4-year contract was awarded in October, 1972.

A preliminary economic analysis has been made for a COED plant with added char gasification. The two major processes of choice from the existing gasification technology are the Koppers-Totzek and Winkler processes. (see separate descriptions in Gasification Section of this report). Estimates for a commercial COED plant have been suggested as follows:

Estimated cost:	U.S. $280–380 million
Coal Usage:	24,000 short tons per day (clean coal)
Oil Production:	26,400 barrels per day
Char Production:	12,000 short tons per day
Gas Production:	390 million S.C.F. per day (400-500 BTU/SCF)

The COED process, since it is based essentially on "here and now" technology, could move into the commercial stage rather quickly, given the right economic climate and with the completion of the char gasification program.

The remaining development work on the COED project consists of performing large-scale char-gasification tests in existing commercial equipment. The first of these char gasification tests will probably be conducted in a Koppers-Totzek gasifier. The program to perform these tests is being sponsored by O.C.R. and the Electric Power Research Institute (EPRI). The FMC Corp. was recently awarded a U.S. $650,000 contract by EPRI to determine whether environmentally-clean gas can be produced economically from char such as that produced in the COED process.

In a report to O.C.R. dated 23/1/75, FMC Corp, detailed the results of successful combustion-performance tests on COED char in experiments conducted in a commercial-size downward-fired boiler owned by the Luzerne Electric Division of U.G.I. Corporation. The test boiler was designed and installed by the Foster-Wheeler Corp., who also participated in the COED char test program and analyzed the boiler performance results.

In other COED-related developments, the COGAS Development Company, a joint-venture partnership of which FMC Corporation is a member, plans to construct a demonstration plant for the COGAS program to process 800–1000 T.P.D. of coal feed. The COGAS process is based on the COED process but with a unique air-steam gasification step to produce pipeline gas or synthesis gas from the char derived from the COED pyrolysis step. (Refer to "COGAS" in this section of the report for further details.)

In a recent report to the Office of Coal Research, the American Oil Co. described the combination of COED pyrolysis with the Molten Salt process of M.W. Kellogg Co. to produce synthetic crude oil and substitute natural gas.

Note: C.S.I.R.O. in Australia are conducting tests on fluidised-bed pyrolysis similar to the COED process utilising a wide range of coals including the Millmerran bituminous coal which is also the object of S.R.C. process study by Mitsui & Co. (See S.R.C.–Mitsui for further details)

C.O.G. (Coal–Oil–Gas Refinery)

Process Developers:

1. Coal liquefaction (S.R.C.) section – Pittsburgh and Midway Mining Co. (a subsidiary of Gulf Oil Corp.)

2. Synthesis gas/hydrogen production by coal gasification – Bituminous Coal Research, Inc. (the research affiliate of the U.S. National Coal Association.)

Sponsor: Office of Coal Research, American Gas Association.

Technical Evaluation Services: The Ralph M. Parsons Company, Los Angeles. Chem Systems, Inc., New York.

Operating Conditions:

Process Section	Temp $^\circ$F	Press.p.s.i.	Reactor Type	Products
S.R.C. *	850	1,000	Extraction-hydrogenation	Naphtha, Fuel Oil, S.R.C. Fuel Gas
BI–GAS *	3,000	200	Slagging, entrained	Synthesis Gas, H_2–Rich Gas

* S.R.C. and BI–GAS processes modified to produce desired products.

Description:

1. Liquefaction Section.

This section includes coal slurrying, dissolution and distillation operations.

The coal feed, crushed to $-1/8$ inch, is combined with unfiltered solvent to produce a 50 wt.% slurry which is sent to the pre-heat furnace. The slurry is combined with synthesis gas and water and fed to the reactor. The resultant mixture from the reactor consists of a liquid phase, a solid phase of ash and unreacted coal, and a gas phase. The gas is separated, scrubbed to remove CO_2 and H_2S, and the major portion is combined with make-up synthesis gas and recycled to the slurry feed. The remaining portion of the gas is utilised as fuel gas. The solid phase is separated by filtration from a portion of the liquid phase and is transferred to the gasification plant where it is gasified to produce the synthesis gas. The rest of the unfiltered solvent product, containing unreacted coal, is recycled to the feed-coal slurry.

The filtrate produced in the filtration step is fractionated to produce a naphtha stream, distillate that is desulphurised to light boiler fuel, and residual fuel oil. Solid solvent-refined coal (S.R.C.) could be produced by distillation if required.

2. Gasification Section.

Wet filter cake from the liquefaction section is introduced into the slagging, entrained-flow gasifier where it reacts with steam and oxygen at elevated temperatures and pressures. The offgas is a carbon monoxide-hydrogen synthesis gas which is sent to heat-recovery, coarse-char cyclones and acid-gas removal.

Most of the clean synthesis gas is sent to the liquefaction section. That portion of the gas not sent to the liquefaction section is converted to high-purity hydrogen that is utilised for the hydro-treatment of the naphtha and fuel oil. The high-purity hydrogen is produced by subjecting the synthesis gas to shift conversion, CO_2 removal and methanation.

The liquefaction and gasification processes can be modified to produce a wide range of products e.g. solvent-refined coal, coal-derived "light refinery liquid", pipeline gas, liquid petroleum gas, synthesis gas for ammonia, hydrogen etc.

Status:

The Ralph M.Parsons Co. has been awarded a U.S. $2.99 million contract by the Office of Coal Research to evaluate the COG refinery concept over three years. The company has completed a preliminary design and capital cost estimate of a demonstration plant to process 10,000 T.P.D. of coal to produce:

Naphtha	— 2,011 barrels per day
Fuel Oil Product	— 8,472 barrels per day
Heavy Liquid Product	— 14,345 barrels per day
Heavy Liquid Plant Fuel	— 121 Tons per day
Fuel Gas	— 2,142 Tons per day
Sulphur	— 317 Tons per day

The design data were supplied by Pittsburgh & Midway Coal Mining Co. for the S.R.C. section and Bituminous Coal Research, Inc. for the gasification section.

Other conceptual designs for the COG plant have been assembled by O.C.R. For "Project Independence", an O.C.R. blueprint team prepared a conceptual liquefaction-gasification plant with input and output balance as follows:

INPUT		OUTPUT	
Coal:	60,000 T.P.D.	Liquid Boiler Fuel:	100,000 B.P.D.
Oxygen:	12,000 T.P.D.	High BTU gas:	580 million SCF/day
Water:	110 acre feet per day	Naphtha:	1,620 T.P.D.
Power:	300 MW.		

Projected selling prices for the products have been estimated at U.S. $1.22 per million BTU. (coal at U.S. $7.25 a ton.)

The rationale for the seemingly premature design of the COG demonstration plant is expedience. Objectives include:

1. Reducing the lead-time for operation of commercial-scale plants.

2. By-passing the pilot-plant programs to gain time for development of commercial processes.

3. Providing a basis for predicting the economics of commercial plants.

4. Defining performance requirements and financial incentives for rapid development of large-scale coal-conversion plant equipment.

5. Demonstrating operability of commercial-scale equipment.

Detailed demonstration-plant design could be in motion during pilot-plant operation.

Chem Systems, Inc., in 1971, working as a sub-contractor to Pittsburgh and Midway Coal Mining Co., also suggested the combination of the S.R.C. and BI—GAS processes to form a COG refinery. Such a refinery would require 57,000 T.P.D. of coal feed and 7,740 T.P.D. of oxygen and from these would produce 1,980 tons of L.P.G., 7,660 tons of pipeline gas, 1,800 tons of sulphur, 14,660 of light refinery liquid fuel oil, 8,850 tons of Solvent Refined Coal (SRC), which is used to run the refinery, and 2,500 tons of SRC for sale, per day. Overall efficiency of the plant would be 75.5%. Investment would total U.S. $786 million.

COGAS

Process Developer:

Cogas Development Company, Princeton, N.J., a joint-venture partnership of:

Consolidated Natural Gas Company
FMC Corporation
Panhandle Eastern Pipe Line Company
Republic Steel Corporation
Tenneco, Inc., through its subsidiary, Tennessee Gas Pipeline Company
Union Pacific Corp., through its sub. Rocky Mountain Energy Company

Technical Services:

Bechtel Corporation

Description:

The COGAS process converts coal into both oil and gas products and to achieve this the gasification-combustion step is integrated with a multi-stage fluidised-bed coal pyrolysis step. The products of the pyrolysis step are a reactive char and pyrolysis oil and gas.

The resultant pyrolysis gas is stripped of light hydrocarbons, and processed along with the synthesis gas from the char gasification. The light hydrocarbons can be produced as a separate stream or blended back to increase the heating value of the product gas.

The char product of pyrolysis is sent to the gasifier while the product oil may be upgraded by hydrogenation to a high-grade synthetic crude oil, or by using less hydrogen in this step, to a low-sulphur fuel oil. The hydrogen for this upgrading is produced by reforming a portion of the product gas.

The product gas stream is suitable for shift conversion, purification and methanation to pipeline quality. For medium—BTU gas production, the synthesis gas is raised to a minimum pressure and cleaned to reduce sulphur and particulates. The resultant clean gas is suitable for power generation or reducing-gas process utilisation such as ammonia or methanol synthesis.

Operating Conditions:

Pyrolysis Section	Temp.^{o}F	Press.p.s.i.	Products
Fluidised Beds (4)	600-1500	5 - 10	Char, Oil, Liquor, Gas
Hydrotreatment Section	**Temp.^{o}F**	**Press.p.s.i.**	**Product**
Catalytic Fixed-bed	750	2500-3100	Synthetic Crude Oil
Gasification Section	**Temp.^{o}F**	**Press.p.s.i.**	**Product**
Gasifier-Combustor	1500-1700	0 - 30	Synthesis Gas

Status:

The COGAS Development Company was formed in 1972 as a Joint Venture Partnership dedicated to the development of the COGAS process. In all, the company has budgeted U.S. $7 million for the developmental phases of its program.

The initial efforts of the company were devoted to laboratory experimentation of the COGAS process, cold model studies simulating pilot-plant operations, and engineering and design for two pilot-plant facilities.

The process is largely based on the COED process of the FMC Corporation which has been under evaluation in a 36 T.P.D. pilot plant at Princeton, N.J. since 1970. The COED development will provide the data for the pyrolysis section of the COGAS process, thus the thrust of COGAS process development is concerned with the gasification of the char derived from the pyrolysis step in a novel route that employs low-pressure gasification using air instead of oxygen to produce, pipeline-quality gas or synthesis gas.

Two approaches to the char gasification—combustion step of the process were piloted because both appeared to offer advantages. For both, half-scale cold models were constructed and operated to study process flow variables before finalising the pilot plant designs.

Based on a design by COGAS Development Co. for one unit, and a design by British Coal Utilisation Research Association (BCURA) for the other, Bechtel, Inc. engineered and constructed two pilot plants.

1. **The Inert Heat-Carrier Pilot Plant.**

    This pilot plant, located in Princeton, New Jersey, was designed by the COGAS Development Co. and has a design capacity of 2.5 T.P.D. of char feed. Construction of the plant was completed in February, 1974 with start-up in May, 1974. This system utilises an inert heat-carrier such as a ceramic or a pelletised coal ash. Char fines for fuel are collected from the product gas stream. Char combustion reheats the fluidised heat-carrier in a combustion vessel which is then transported back to the gasifier where it showers through the bed of fluidised char which reacts with steam to produce a synthesis gas. The combustor which heats the inert heat-carrier is air-blown.

    Operation of this pilot plant is continuing.

2. **The Char Heat-Carrier Pilot Plant;**

    This pilot plant was designed by BCURA and is located at the BCURA laboratories in Leatherhead, England. The plant achieved initial start-up operations in April, 1974. Design capacity of this pilot plant is 50 tons per day of char feed.

    This system utilises circulating char as the heat-carrier combined with slagging combustion to facilitate ash separation with minimum carbon loss. Char from the pyrolysis section of the COGAS process is fed to the gasifier where it is gasified with steam utilising heat supplied by recycled hot char from the air-blown combustor. Char fines separated from the flue-gas stream are fed to the combustor. Coarse char is separated from the product-gas stream and is recycled to the lift tube where the hot combustion gases reheat it for recycling to the gasifier.

    Operation of this pilot plant is continuing.

Since initial start-up, the pilot plants have successfully demonstrated most of the key features of the systems.

An integral part of the overall COGAS program and one which is proceeding concurrently with the pilot plant program, is an extensive process engineering study being conducted with the assistance of Bechtel Associates Professional Corporation, New York. This study is intended to oversee all of the steps in the COGAS process and to achieve their optimisation.

COGAS Development Company is currently planning for a demonstration plant with a feed capacity of 800 − 1,000 T.P.D. of coal.

CONSOL SYNTHETIC FUEL (C.S.F.)
(Previously known as "PROJECT GASOLINE")

Process Developer: Consolidation Coal Company, a subsidiary of Continental Oil Co.

Sponsor: Office of Coal Research
American Electric Power Company
Allegheny Power System, Inc.

Description: Coal, crushed and dried, is preheated in a fluidised-bed to $450^{O}F$ then slurried in a process-derived solvent. The slurry is pumped to a stirred, solvent-extraction vessel where extraction occurs at $765^{O}F$ and 150-400 p.s.i. Vapours produced in the extraction vessel are sent to the fractionation and solvent-recovery section. Liquid product and solid residue, consisting of unreacted coal and mineral, are separated in hydrocyclones. Liquid is sent to the solvent-recovery and fractionation section where it is combined for treatment with the extraction vapours. Recycle solvent, distillate, and liquid product for hydro-treatment to naphtha are produced in this section.

Solid residue in a concentrated slurry is sent to a low-temperature carbonisation reactor where the slurry is pyrolysed at $800 - 900^{O}F$ and low pressure to recover the solvent and also to produce additional distillate by coking the residue and the extract. Overhead vapours are quenched and oil scrubbed to recover the condensables from the fixed gas. This produces a light liquid product and a clean fuel gas.

Char product from the carbonisation reactor is used in a steam-oxygen reactor for hydrogen generation which is employed in the hydrotreatment and distillation step.

In the hydrotreatment and distillation section liquid product from the fractionation step and tar from the carbonisation step are hydrogenated in a catalytic reactor operating at $800^{O}F$ and 3,000 p.s.i. to produce naphtha, light oil, gas and make-up solvent. Residue from this step can be used as plant fuel.

Operating Conditions:

Reactor	Temp.^{O}F	Press.p.s.i.	Products
Extraction	765	150−400	Vapours, Liquid, Residue
Carbonisation	925	10	Char, Pyrolysis Liquid and Gas
Hydrotreatment	800	3,000	Naphtha, Oil, Gas, Residue

Status: The Office of Coal Research and Consolidation Coal Company entered into a contract in September, 1963 for the development of the C.S.F. process. A 20 T.P.D. pilot plant was designed utilising data obtained from extensive bench-scale testing by Consolidation Coal and construction of the pilot plant was completed in May 1967, at Cresap, W. Va. This plant also included a 13 T.P.D. hydrogenation plant for the conversion of coal extract into low-sulphur distillates.

The results of the pilot-plant runs were less favourable than those predicted by bench-scale testing, particularly in hydrogen consumption and catalyst life. The plant was put on a care-and-maintainance basis after April, 1970 to allow for a period of reassessment of the process and to resolve the mechanical problems encountered in the operation of the plant. A detailed engineering survey of the plant was conducted by Foster-Wheeler Corp. in 1970 who reported to O.C.R. that the process was technically sound but that major mechanical modifications were necessary to increase operating reliability.

Standard Oil of Ohio (SOHIO) held some of the background patents for the process and has joined with Consolidation Coal for the design of a 900 T.P.D. C.F.C. process demonstration plant at Toledo, Ohio, which will draw upon the results obtained from the Cresap pilot plant for design data for elements of the C.F.C. plant. (see S.R.C.–C.F.C. in this section of the report).

Late 1974, O.C.R. announced that the Fluor Corporation would undertake a three-year project aimed at revamping the Cresap pilot plant. Under the new contract, Fluor will supplement the existing plant equipment with new equipment designed to produce fuel oil for power generation and industrial uses from coal.

The Cresap plant may also be modified to evaluate several coal liquefaction processes for various types of coal. Elements of the H-COAL process may be incorporated in the revamped plant.

The contract will be funded U.S. $11.3 million by O.C.R., with American Electric Power Company and Allegheny Power System, Inc. contributing U.S. $1 million. These two power utilities are believed to be interested in evaluating the utilisation of coal-derived, low-sulphur, liquid and solid fuels for power generation to overcome sulphur-emission controls applied by Federal air-quality rules, seeing the solution to the problem being an efficient front-end process rather than power-plant stack controls.

Estimated yields from a conceptual C.S.F. commercial plant include 34,200 barrels per day of heavy fuel oil, 13,200 barrels per day of naphtha, and 170 million S.C.F. per day of 820 BTU/SCF gas from a coal feed rate of 29,000 tons per day.

COSTEAM

Process Developer: U.S. Bureau of Mines, Pittsburgh Energy Research Centre, Pittsburgh, Pa.

Description: A slurry of pulverised coal in some of the product oil is pumped with carbon monoxide, or a CO-rich synthesis gas, into a stirred reactor at $800^{\circ}F$ and 4,000 p.s.i. The steam for the reaction is derived from the moisture of the coal. Products from the reactor go to a receiver where the raw oil is separated from the product gas. Unreacted coal and mineral are removed from the product oil by a centrifuge or a pre-coat filter.

The process is specifically designed for coals with high reactivities and high moisture contents, such as lignite. With highly-reactive low-rank coals, no catalyst is required in the process.

Operating Conditions:

Reactor	Temp.$^{\circ}$F	Press.p.s.i.	Reactants	Products
Stirred extraction	800	4,000	Coal-CO-Steam	Fuel oil, Gas

Status: The Bureau of Mines has investigated the process in batch-autoclave studies in a 2.25 gallon vessel.

Results to date have been encouraging. The Bureau plans to continue testing of the process to provide sufficient data for the design and fabrication of a pilot plant.

EXTRACTIVE–COKING PROCESS

Process Developer: Arthur D. Little, Inc., Acorn Park, Cambridge, Mass.

Sponsor: Office of Coal Research

Description: This process is a modification of the delayed-coking process. Bituminous coal is subjected to extractive coking to produce low-sulphur liquid fuels such as No. 2 fuel oil, a high-quality residential heating fuel.

No details are available on process routes or operating conditions.

Status: The Office of Coal Research recently awarded a U.S. $568,000 contract to Arthur D. Little, Inc. to make an experimental study of the process.

The major objective is to demonstrate the technical and preliminary economic feasibility of the extractive-coking process utilising a 20-40 lb. batch extractive coker at the Foster-Wheeler laboratories, Livingston, N.J.

Arthur D. Little also won a contract in June, 1974 to assist the State of Illinois in the development of its coal resources.

EXXON LIQUEFACTION

Process Developers: Exxon Research and Engineering Co., Baytown, Texas, and Carter Oil Co., both subsidiaries of the Exxon Corp.

Description: Crushed coal is slurried with a recycle solvent, preheated to about 800°F, and then pumped into the liquefaction reactor operating at about 2,000 p.s.i. Preheated hydrogen is also added to the reactor. The product from the liquefaction reactor is sent to the separation step where gas, naphtha, recycle solvent, distillate, and heavy bottoms (containing unconverted coal and minerals) are separated by distillation.

The recycle solvent is catalytically hydrogenated and is then slurried with fresh coal feed to repeat the cycle.

The raw liquid product is upgraded by hydrogenation to adjust the quality of the product by desulphurisation and conversion.

Finally, the heavy bottoms and solid residue go to a gasifier for hydrogen production.

Operating Conditions:

Reactor	Temp°F	Press. p.s.i.	Reactants	Products
Catalytic Hydrogenation	800	2,000	Coal–H_2–H–Donor Solvent	Naphtha, Gas, Distillate, Heavy Bottoms, Recycle Solvent

Status: The Synthetic Fuels Research Division of Exxon Research and Engineering Company have been developing this process in several small continuous pilot plants up to 0.5 ton per day in size. General process feasibility has been established with bituminous and sub-bituminous coals.

The design of a 200 T.P.D. pilot plant, to be located adjacent to Exxon's Baytown, Texas, refinery is currently underway. A pilot plant of this size would demonstrate critical process features for scale-up to a commercial-size plant.

Exxon is seeking federal government financial support for the construction and operation of the 200 T.P.D. plant.

14

FISCHER–TROPSCH SYNTHESIS

Process Developers: Fischer and Tropsch (Germany) M.W. Kellogg Co. (Synthol Process, SASOL)
Arge-Arbeit Gemeinschaft Lurgi and Ruhrchemie (Arge Synthesis, SASOL)

Description: As only one commercial plant, SASOL in South Africa, is currently in operation to produce liquid hydrocarbons from coal-derived synthesis gas via Fischer-Tropsch Synthesis, the following description is for that plant:

Coal is gasified in a battery of 13 Lurgi high-pressure, steam-oxygen gasifiers to produce a gas consisting essentially of carbon monoxide and hydrogen, with a proportion of other gases, tar and oil. The gas stream from the gasifiers is quenched to remove tar and oil and purified by the Rectisol (Lurgi) process which uses a single solvent (methanol) to remove the last traces of tar and oil, carbon dioxide, hydrogen sulphide, organic sulphur, ammonia and phenol. The purified synthesis gas stream is partitioned and a part of the gas is passed through a fixed-bed catalytic reactor (Arge synthesis). Feed gas has a H_2/CO ratio of about 2:1 and synthesis occurs under conditions of $430^{\circ}F$ and 360 p.s.i.

The products of the Arge synthesis are straight–chain, high–boiling hydrocarbons, with some medium–boiling oils, diesel oil, L.P.G., and oxygenated compounds such as alcohols.

The portion of the synthesis gas which was not sent to the Arge unit goes to the Synthol plant (Kellogg synthesis) which is a fluidised-bed catalytic (iron) reactor. In this reactor, catalyst is circulated along with the synthesis gas. Gas and catalyst leaving the reactor are separated in cyclones and the catalyst is recycled. Operating conditions are 600°– $625^{\circ}F$. and 330 p.s.i.

A portion of the Synthol plant tail gas is reformed with steam to increase the H_2/CO ratio to about 3:1, and is mixed with the fresh synthesis gas.

Reactor effluent gas is quenched in a scrubbing tower where the remaining catalyst dust is removed and returned to the reaction zone in the form of a heavy slurry oil.

The raw products from the synthesis require certain treatment and then final purification to make the specification products. From the gas phase, valuable hydrocarbon and chemical products are scrubbed out and recovered. The oil phase is treated catalytically to remove dissolved oxygenates and then distilled into gasoline and fuel oil fractions. The remaining liquor is distilled and fractionated to produce chemical products. Heavy alcohols to pentanol are also recovered.

A portion of tail gas from the Arge and Synthol synthesis plants is removed and used for utility gas.

Operating Conditions:

Synthesis Process	Catalyst	Temp.$^{\circ}$F	Press p.s.i.	Products
Arge Fixed Bed	Iron/Cobalt	450	360	Petrol, L.P.G., Oil, Wax, Gas
Kellogg Fluidised Bed	Iron	620	330	Petrol, Alcohol, Oil, Gas

Status: Fischer and Tropsch made extensive studies of the catalytic reduction of carbon monoxide to various hydrocarbon liquids during the period 1923-1933. Following pilot plant operations in 1932 a 1,000 tpy plant was constructed by Ruhrchemie in 1933 and a 30,000 tpy plant in 1936. By 1939, nine plants had been erected in Germany and one in France. The production of gasoline, diesel and paraffins by Fischer-Tropsch synthesis, along with the Bergius process products, provided a significant contribution to the energy requirements of war-time Germany.

As soon as the process was commercially available in 1935, a South African concern, sponsored by the Anglo-Transvaal Consolidated Investment Co. Ltd., acquired the rights on the process for the Union of South Africa. In 1947, the Liquid Fuel and Oil Advisory Board was established. A committee established by the Advisory Board recommended that the available Fischer-Tropsch type processes should be commercially viable in the South African situation and that rights to coal deposits, land and process should be acquired from the Anglo—Transvaal Investment Co. Ltd. by a government—financed company.

The South African Coal, Oil and Gas Corporation Ltd. (SASOL) was incorporated under the Companies Act in September, 1950 with all its shares being held by the Industrial Development Corporation, a government company with its own charter.

SASOL invited several internationally-known companies to submit proposals for a coal-based Fischer-Tropsch synthesis plant. The successful tenderers were M.W. Kellogg Co. (U.S.A.) with the Synthol process, and Arge-Arbeit Gemeinschaft Lurgi and Ruhrchemie with the Arge synthesis.

In 1954, 10 Lurgi gasifiers were installed at Sasolburg to produce the synthesis gas for the Arge and Kellogg F-T plants. In 1955, the first gasoline was produced from Transvaal coal.

In 1966, three additional Lurgi gasifiers were added to the Sasolburg plant to increase production. SASOL currently provides feedstock for three refinery plants, in Capetown, Durban, and in Sasolburg.

In December 1974, the South African government released data on a second Fischer-Tropsch plant to be constructed by SASOL on the Eastern Transvaal coalfields. This plant is expected to cost U.S. $1.5 billion with foundations expected to be laid in April 1977 and to come on-steam 1979-1981. The SASOL 2 complex will employ only a refined Kellogg Synthol process for F-T synthesis of Lurgi-produced gas, unlike the Sasolburg plant which also uses the Arge synthesis to produce waxes and other heavy products.

SASOL 2 will produce mainly gasoline and fuel oil, although there will be other products including an estimated 150,000-200,000 ton per year of ethylene, 250 ton per day of ammonia, 92,000 ton per year of sulphur and 190,000 ton per year of tar products.

Sasolburg's current output is 2.5×10^5 ton per year of petrochemicals which includes 1.68×10^5 ton per year of gasoline. SASOL 2 should produce something of the order of 1.5 million ton per year of gasoline. Thus SASOL 2 should represent a 9-times scale-up of SASOL I at Sasolburg.

The SASOL 2 plant will give the company 30 per cent of the nation's automotive fuels market in 1980.

In the U.S.A., interest in Fischer-Tropsch synthesis is increasing. In 1974, the Office of Coal Research awarded the Ralph M. Parson's Company of Pasadena, Calif., a three-year, U.S. $2.99 million contract to evaluate the COG refinery concept. This contract entails preliminary design work on a Fischer-Tropsch synthesis plant to produce substitute natural gas and motor fuels as the main products.

The U.S. Bureau of Mines and the Office of Coal Research have a Fischer-Tropsch research and development program underway. Preliminary design for a process development unit to evaluate modifications of the process is in progress with operation of the unit scheduled for 1977.

The interest by the Bureau of Mines in Fischer—Tropsch studies dates from the time when the Office of Synthetic Liquid Fuels was set up in 1944. Coal research and liquefaction plants where established over the following eleven years in what are now the Pittsburgh and Bruceton Energy Research Centres, where F—T synthesis was evaluated over a number of years. At Louisiana, Mo., a government ammonia plant was converted into two coal liquefaction demonstration plants. One utilised the Bergius process and the other employed the F–T synthesis to produce about 100 barrels per day of gasoline and chemicals.

Blaw-Knox provided the detailed design, procurement, and erection of the Bureau's Fischer-Tropsch Synthesis Pilot Plant located in Bruceton, Pa. during the time of the Office of Synthetic Liquid Fuels.

Note: Despite the often-expressed opinion that the F-T process at Sasolburg is heavily dependent on government subsidy, the 1973–74 report shows from sales of R 293 million a pre-tax profit of R17 million was realised.

GARRETT'S COAL PYROLYSIS

Process Developers: Garrett Research & Development Co., La Verne, Calif., a subsidiary of Occidental Petroleum Corp.

Sponsor: Occidental Petroleum Corp.

Description: Crushed coal is introduced into a pyrolysis vessel in a stream of recycled gas where it is pyrolysed at 1100°F through contact with hot char from the char heater vessel. Effluent from the pyrolysis reactor is sent to a series of cyclones for separation of gas and entrained char. A portion of the char is cooled as product char. The remaining char is sent to the char heater where a portion is burned with air to heat the contents to 1400°F. Hot char is cycled to the pyrolysis reactor.

The gas from the cyclones is cooled and scrubbed, tar is removed in a gas-liquid separator and after acid-gas removal, a portion of the gas is separated as product gas (700 BTU/SCF) to be utilised as fuel gas or upgraded to pipeline quality by methanation. The remaining portion of the gas is sent to a hydrogen plant for generation of hydrogen which is used in the tar hydro-treater for the production of liquid products.

Operating Conditions:

Reactor	Temp.$^{\circ}$F	Pressure	Reactants	Products
Pyrolyzer	1100	Atomospheric	Coal-Hot Char	Tar, Product char, Gas (700 BTU/SCF)
Char-Heater	1200-1600	Atmospheric	Char-Air	Hot char for Pyrolyzer

Status: Garrett Research and Development Co. has been working on the pyrolysis route to coal liquefaction at the La Verne plant for some years. Early evaluation of possible processes indicated to Garrett that pyrolysis provided the simplest and lowest-cost method for rapid commercialisation of coal liquefaction. Development work was initiated with a review of existing pyrolysis data which determined the most attractive route for an efficient conversion would incorporate high heat transfer and short residence time in the reactor. Mathematical model studies of the coal-heating process, with the emphasis on short-residence time parameters, demonstrated that pulverized coal can be heated to decomposition temperatures in fractions of a second. These theoretical results were confirmed in tests on high-volatile bituminous coals in a small, semi-continuous reactor. Results indicated a maximum conversion of coal to liquids of 35% at 1075°F. which is roughly twice the yield predicted by Fischer assay and significantly higher than fluidised-bed pyrolysis. The average yields were 56.7% char, 35% tar, 6.5% gas and 1.8% water.

The process is currently being tested in a 3.6 T.P.D. pilot plant at the La Verne plant.

(Refer to "Garrett's Coal Gasification" in Section II of this report for additional information).

GAS EXTRACTION

Process Developer: Coal Research Establishment, Stoke Orchard, United Kingdom.

Sponsor: National Coal Board of the United Kingdom.

Description: Pulverised coal is treated with compressed gases at temperatures of 350°-400°F, causing a portion of the coal to go into solution in the compressed gas. The mineral matter (ash) and undissolved coal remain when the coal solution is transferred to a vessel at lower pressure where the refined coal extract is precipitated and the compressed gas released for recompression and recycling to the extraction vessel. The coal extract produced by this process is richer in hydrogen than that produced by liquid-solvent extraction and is thus more suitable for the production of hydrocarbon oils and chemicals.

Status: At Stoke Orchard, there are experimental gas-extraction units for the preparation and recovery of coal extract, for investigation of the processes controlling the rate of extraction to obtain design data for a pilot plant, and a unit to investigate extraction in the presence of reducing gases such as hydrogen, or carbon monoxide and steam.

Results have indicated that this process affords significant advantages over liquid-solvent extraction:

1.   Filtering to remove insoluble residue is not necessary.

2.   Recovery of gaseous solvent is virtually complete.

3.   The extraction residue is a porous solid which is suitable for gasification.

4.   More mobile liquids of higher hydrogen content are obtained than with liquid solvents. (See also "SOLVENT EXTRACTION — C.R.E." in this section of the report.)

H–COAL

Process Developer: Hydrocarbon Research, Inc., Trenton, N.J., a subsidiary of Dynalectron Corporation.

Sponsors: Office of Coal Research
Atlantic Richfield Corp.
Ashland Oil Co.
Contental Oil Co. (through Consolidation Coal Co.)
Gulf Oil Corp. (through Gulf Mineral Resources)
Esso (through Carter Oil Company)
Standard Oil of Indiana (AMOCO)
Sun Oil Company
Electric Power Research Institute.

Description: Dried, pulverised coal is slurried with process-derived oil to which is added compressed, make-up hydrogen. The slurry and hydrogen are heated in a fired preheater before being fed to the base of the catalytic (Co/Mo) ebullated-bed reactor, along with recycled high-pressure gas that is similarly preheated. The catalyst in the reactor is suspended in the ebullated-bed by the additional internal recycle-oil flow provided by the hot-oil recycle pump. The relative sizes of the catalyst and coal is such that only the unconverted coal, ash, liquid and gaseous products leave the reactor, while retaining the catalyst therein. The reactor product slurry is let down at essentially reactor temperature to the atmospheric pressure flash drum in which a portion of the lighter hydrocarbon liquids is flash-vapourised and fed to the atmospheric distillation tower. The slurry material remaining after the atmospheric flash can be further flashed in the vacuum drum to obtain vacuum distillate overhead and a vacuum bottoms slurry product.

The reactor vapours not condensed at ambient temperatures are scrubbed in the recycle-gas scrubber. Light hydrocarbons are removed and the hydrogen concentration is increased in this scrubbing systems before the gas is returned to the reactor.

The oil used for slurry preparation could be a combination of several streams derived from the catalytic reactor, the vacuum distillate fraction, the atmospheric bottoms fraction or hydroclone clarified liquor, depending on slurry oil composition desired.

Operating Conditions:	Reactor	Temp $^\circ F$	Press. p.s.i.	Reactants	Products
	Catalytic, ebullated bed	850	2,250-2,700	Coal-oil-H_2	Synthetic Crude Oil, Gas

Status:

The H-COAL process was developed by Hydrocarbon Research, Inc. as a further application of the catalytic, ebullated-bed reactor technology employed to convert heavy oil residues into lighter fractions by the H–OIL process originally developed by HRI and Cities Services Oil Company.

For the past eleven years, HRI has been developing the H–COAL process, first in a 25 lb./day bench-scale unit, then in a Process Development Unit (PDU) handling 3 T.P.D. of coal.

In February, 1965, the Office of Coal Research awarded a contract to HRI for the bench-scale and PDU operations and preparation of a conceptual process design. Owing to lack of funds, the contract was cancelled in September, 1967. Subsequent to this, HRI continued research and development alone until Atlantic Richfield began funding the project in 1968. The support continued until 1970. During the period 1971-72, the program was supported by a consortium of six companies: Ashland Oil Co. Atlantic Richfield, Consolidation Coal Co., Esso Research & Engineering Co., Gulf Mineral Resources, and Sun Oil Co. In 1973, three companies left the consortium for budgetary reasons, and Standard Oil of Indiana entered the program.

In September, 1974, O.C.R. awarded a U.S. $8.1 million, 14-month contract to H.R.I. for the engineering portion of a multiphase project for construction and operation of a 600 T.P.D. pilot plant. Design specifications call for the pilot plant to produce low-sulphur fuel oil, and synthetic crude suitable for refinery processing into gasoline, kerosene, diesel fuel, fuel oil and petrochemical feedstocks.

The engineering design will be performed by the Hydrocarbon Groups engineering subsidiary, HRI Technical Services, Inc., Miami, Fla. Experimental work will continue at HRI's Trenton, N.J. laboratory where the process has completed over 4,000 hours of testing in the 3 T.P.D. P.D.U. and over 46,000 hours of bench-scale testing at from 25 to 100 lb./day. A recent bench-scale run attained 2,700 lb. of coal per pound of Co/Mo catalyst.

March, 1975, O.C.R. announced that the 600 T.P.D. pilot plant would be located in Catlettsburg, Kentucky.

The total cost of the completed 600 T.P.D. pilot plant will exceed U.S. $80 million, with O.C.R. providing two-thirds of the funds and an industry team providing one-third. The industry team includes Ashland Oil Co., Atlantic Richfield, Standard Oil of Indiana, Sun Oil Co. and Electric Power Research Institute (EPRI.)

Elements of the H–COAL process are to be included in the revamped Cresap, W.Va., C.S.F. pilot plant under a recent O.C.R. contract awarded to Fluor, Inc.

Dow Chemical Co. is currently evaluating the suitability of the H–COAL process, along with COED and SRC processes, for commercial-scale application under another OCR contract.

HRI's concept of a commercial plant processing 25,000 T.P.D. of coal calls for a plant investment cost of U.S. $370 million with annual operating costs of U.S. $196 million (1972 costs). The plant would produce crude oil, fuel gas, sulphur and ammonia.

The cost of hydrogen utilised in the process will be a significant factor in the overall economics of a commercial plant.

In another H–COAL–related development, O.C.R. awarded a 16-month, U.S. $201,023 contract to Stanford Research Institute, Menlo Park, Calif. on 2nd January, 1974 to determine the economic feasibility of a coal-conversion plant in south-western Alaska utilising the large Beluga sub-bituminous coal reserves. Two processes, H–COAL and Solvent-Refined Coal (S.R.C.), will be evaluated for suitability. The products from an S.R.C. or H–COAL plant would be suitable for export to the Pacific-rim countries, notably Japan and the United States. The study will also estimate the landed price and examine in detail the markets for these products in Japan and the U.S. west coast. This study is funded by O.C.R. (U.S. $153,023) and jointly by Placer AMEX and Nissho–Iwai American, a subsidiary of Nissho–Iwai of Japan (U.S. $48,000).

The results of this preliminary study should be outlined in O.C.R.'s 1974-75 Annual Report.

HYDRO–CRACKING PROCESS

Process Developer:	Continental Oil Company.
Sponsor:	Office of Coal Research.
Description:	No details are available on this process.
Status:	The Office of Coal Research has recently awarded the Continental Oil Company a four-year, U.S. $6,450,335 contract to determine the technical and economic potential of the Hydro-Cracking process for the production of distillate fuels from coal.

The research and development program will be conducted by Continental Oil Co.'s subsidiary, Conoco Coal Development Division at their Library, Pa. research centre utilising a bench-scale continuous hydro-cracking unit and a fluidised-bed combustion unit for regeneration of zinc halide catalyst in the first stage of the program.

The program's second phase involves construction and operation of 100 lb/hour P.D.U.'s. Conoco and Shell Development Corp. are providing funds of U.S. $800,000 and O.C.R. U.S. $5,650,335.

The goal of the program is to refine the process to produce about four barrels of oil per ton of D.A.F. coal.

INTERMEDIATE HYDROGENATION

Process Developer:	University of Utah, Dept. of Fuels Engineering, Salt Lake City, Utah.

Sponsors:	The State of Utah.
	Office of Coal Research.

Description: This coal conversion process depends upon the application of stannous or zinc chloride as a catalyst in a quantity of about 5 per cent of the weight of the coal. Dry, pulverized coal is passed with the catalyst into a reactor containing a large excess of hydrogen at 2,000 to 2,500 p.s.i. Residence time in the reactor is of the order of seconds during which time hydrogenation of the coal to produce liquid and gaseous products occurs.

A preheater is employed to bring the coal-hydrogen mixture to a temperature just below the plastic range of the coal at the entrance of the reactor. This ensures the coal particles will reach reaction temperatures (930–1020°F) within the length of the reaction tube.

The chloride catalyst is essentially completely recovered from the reaction products for recycling.

Operating Conditions:

Reactor	Temp.$^{\circ}$F	Press.p.s.i.	Reactants	Products
Catalytic entrained-flow	930-1020	2,000-2,500	Coal-H$_2$	Liquid and Gas

Status: Early work at the University of Utah included bench-scale testing of the process. On 2nd October, 1969, O.C.R. awarded a 5-year, U.S. $1,608,434 contract to the University of Utah to develop the process. Of this total, O.C.R. supplied U.S. $1,335,000 and the State of Utah provided the remaining U.S. $273,434. This contract called for the operation of a 50 lb. per hour continuous process unit to obtain data and operational experience to make well-founded predictions and evaluations concerning the process on a larger scale.

Tests at design rate in a 6 ft. long by 0.5 inch diameter reactor were completed by the expiry date of the contract in October, 1974. All project goals were attained including the gathering of design data for fabrication and operation of a process development unit of 5–10 T.P.D. coal feed rate.

The University of Utah has also developed a process for converting coal-derived synthetic oil from the Intermediate Hydrogenation process into BTX, paraffinic naphtha and a product gas. This second-stage process employs a combination of hydro-refining, hydro-cracking and solvent extraction. Approximately 30% of the coal is converted to gas in the process.

Other areas of investigation by the University lies in bench-scale experiments on catalytic hydrosolvation of coal, primarily lignite, to low-sulphur synthetic crude oil in a single step, coal pyrolysis, dispersing of coal in a solvent and solution of coal in the presence of ultra-sonic radiation.

LEAS LIQUEFACTION

Process Developer: Leas Bros. Development Corp.

Description: The process employs a single, three-stage reactor tower for the simultaneous production of liquid and gaseous fuels from coal. Crushed, dried coal is mixed with a solvent or recycle oil and the resultant slurry is fed to the top of the reactor tower where it is mixed with silica sand or metallurgical slag and a hydro-cracking catalyst. The coal liquids are removed from the top zone and hydro-cracked with hydrogen or synthesis gas introduced to the bottom of the zone before being sent to the fractionation unit. There the residual heavy fraction is coked by utilising heat generated in the middle zone of the reactor tower where coke, sand and other carbonaceous materials are gasified to yield a producer gas, which is taken off the top of the tower and sent to the fractionation unit. Fly ash is fluidised and removed in the middle zone. The solids flow downward to the bottom zone of the tower where remaining carbon is gasified with carbon dioxide and oxygen to produce carbon monoxide. The carbon monoxide stream is desulphurised then sent to the top stage to produce hydrogen and to supply the carbon dioxide for the bottom stage. Sulphur gases are removed from gas streams and passed through a Claus unit for recovery of elemental sulphur. The liquid and gaseous products are thus relatively free of sulphur.

Status: U.S. Patent 3,779,893 of 18th December, 1973 has been assigned to Leas Bros. Development Corp. This patent describes the above process. Leas Bros. have tested the process on Illinois and Western coals and results indicated that conversion to liquid and gaseous products of D.A.F. coal ranged from 88–91%. No details are available on the current status of development work on the process.

LURGI–RUHRGAS

Process Developers: Lurgi GmbH, a subsidiary of Metallgesellschaft AG, Frankfurt, and Ruhrgas AG.

Description: Feed coal and a heat carrier consisting of hot char are continuously supplied to a mechanical mixer which ensures a uniform mixing of the two components as well as a very rapid equalisation of temperature between the char and coal so that a major part of the carbonisation occurs at the end of the mixer. The resultant pyrolysis gases and vapours are withdrawn at the end of the mixer, passed through a cyclone for dust removal, and then sent to a condensing unit.

The tar is subjected to dust removal and hydrogenated in the hydro-treatment section to produce a range of liquid products. The gas after cleaning, has a heating value of 700-850 BTU/SCF. This gas may be used as a source of hydrogen for the tar-hydrogenation steps or methanated to pipeline quality.

The char which has been used as a heat carrier and newly-formed char, fall into the carboniser shaft where additional temperature equalisation between the heat carrier and fresh distillation residue takes place so that a subsequent degasification can occur. The char leaves the carboniser shaft at the lower end and flows to a lift pipe where it is raised by combustion gases and heated simultaneously. The combustion gases are produced in the lift pipe itself, into which preheated air is blown to cause partial combustion of the char. Char and combustion gases are separated and the gases, after cleaning, are exhausted.

The hot char is collected in a bin and then recirculated to the mixer to complete the cycle. The continuous production of fresh char results in a surplus of circulating char. This surplus is continuously withdrawn.

The close intermixing of coal and hot char in the mixer avoids the formation of agglomerates so that caking coals can be treated.

Operating Conditions:

Reactor	Temp.°F	Pressure	Reactants	Products
Carboniser	1100	Atmospheric	Coal–Hot Char	Char, Tar, Gas

Status: The process was initially tested in a continuous laboratory unit with results indicating tar yields between 129 and 212 per cent of the yield of tar by Fischer Assay. These encouraging results were confirmed in a 12 ton per day pilot plant, with a variation of tar yields of 137–189 per cent of that predicted by Fischer Assays with coals containing 35—46 per cent volatiles (D.A.F.)

The tar produced in the L—R process contains more high-boiling point fractions than tar produced by a slow carbonisation e.g. Fischer retort. The L—R process, by carrying out the carbonisation in a mixer, ensures rapid heating of the coal and thus short residence times in the reaction (of the order of seconds or even fractions of a second) for both char and resultant pyrolysis vapours. This flash carbonisation suppresses the thermal decomposition of the primary tar within the grains and results in much higher yields of vapours than a slow heating.

Gas yield varies inversely to the tar yield. In the flash-carbonisation of high-volatile bituminous coals, tests indicated that the yield of gas was between 220 and 500 SCF per ton of D.A.F. coal i.e. 60 per cent of that indicated by Fischer Assay. Thus, flash-carbonisation by the L—R process increases tar yield and decreases gas yield.

The first commercial L—R flash-carbonisation plant was erected in 1963 at Lukavac, Yugoslavia, processing 1,600 ton per day of lignite. Since then, numerous large-scale plants have been erected around the world. It is not known to what extent the tar products from these plants are utilised for the production of synthetic oils by hydrogeneation.

METHANOL SYNTHESIS

Description:

Methanol can be synthesised by the catalytic reaction of synthesis gas produced by any one of a number of commercially-available coal gasification processes which produce CO/H_2 mixtures. Suitable gas for the catalytic production of methanol utilising conventional zinc-chromium (high pressure) or copper-zinc-chromium (low to moderate pressures) catalysts is produced by passing the raw gas from the gasifier through a conventional water-gas shift conversion to adjust the H_2:CO ratio in the gas to 2:1. After purification, this $2H_2$:1CO gas is sent to the catalytic reactor operating at about 500°F and 750—4,500 p.s.i. (depending on the catalyst) where methanol is formed by the following reactions:

$$CO + 2H_2 \rightleftharpoons CH_3OH \text{ (methanol)}$$
$$CO_2 + 3H_2 \rightleftharpoons CH_3OH + H_2O.$$

Methanol formed in the catalytic converter is condensed and recycle gas is separated for return to the converter. The raw methanol is distilled for purification and higher alcohols (through pentanol) are recovered as residue together with other organic compounds.

Status:

The commercial-scale production of methanol via the catalytic synthesis of coal-derived synthesis gas has been practised in many countries for a great many years, primarily based on Koppers—Totzek, Lurgi and Winkler gasifiers.

Owing to the increasing interest in the use of methanol either as a clean-burning automotive fuel, or as a substitute for tetra-ethyl lead in conventional gasoline to upgrade the octane rating of the gasoline, various projects are currently underway on coal-based methanol synthesis processes.

The U.S. Bureau of Mines is currently evaluating the synthane gasification process for suitability as a source of synthesis gas for methanol production and is conducting experiments with test automobiles run on methanol. (Refer to "Synthane" process in Section II of this report.)

The Consol Division of Continental Oil Company has displayed interest in the utilisation of methanol as a substitute fuel for gasoline and is currently doing research along this line.

In the past, methanol synthesis was operated at pressures of 3,000 to 4,500 psig utilising zinc oxide/chromium oxide catalysts but recent developments by Lurgi and Imperial Chemical Industries (I.C.I.) have led to the commercial development of highly-active copper catalysts which operate efficiently at pressures from 750 to 1,200 psig.

Various estimates have been prepared on the large-scale conversion of coal to methanol via coal gasification. Generally, it can be expected that the efficiency of the energy recovery of coal-to-methanol is about 40—50%, depending on the individual efficiencies of the gasification and synthesis steps. A commercial-scale plant processing about 15,000 T.P.D. of coal would produce about 7,500 T.P.D. of methanol product.

In the U.S.A., several 250 million SCF/day S.N.G. plants (mainly based on Lurgi gasification) are planned. An equivalent-size plant producing methanol instead of pipeline gas should produce about 100,000 barrels per day of methanol. An alternative commercial concept is of a co-product methanol S.N.G. plant in which a portion of the synthesis gas produced in the gasifiers is subjected to shift-conversion to produce a $2H_2:1CO$ gas for methanol synthesis while the remaining synthesis gas is shift-converted to a $3H_2:1CO$ stream for upgrading to S.N.G. by catalytic methanation. With such a scheme, a coal feed of 15,000 T.P.D. could be converted into 185 million S.C.F./day of pipeline gas and 3,625 ton per day of methanol co-product.

It is anticipated that the demand for methanol will accelerate in the future and that large-scale coal-to-methanol plants will be needed to satisfy this demand.

Note: An O.C.R. contract has recently been awarded to Mobil Oil Corporation which calls for developmental studies on the conversion of methanol (probably derived from the gasification of coal) to gasoline. Under this U.S. $862,322 contract, Mobil will conduct the bench-scale demonstration of the technical feasibility of a catalytic process for the conversion of methanol to high-octane gasoline, and design a pilot plant to convert 100 barrels per day of methanol. The work under this contract will be carried out at Paulsboro, N.J. and at the Mobil Technical Centre, Princeton, N.J

MOLTEN SALT CATALYSIS

Process Developer: Montana State University.

Description: Lignite and sub-bituminous coals are converted to liquid and gaseous products by the catalytic reaction of the coal with carbon monoxide and steam at temperatures from 750° to 900^oF. and a final pressure of 5,500 p.s.i. Large amounts of molten potassium and zinc chloride mixtures catalyse the liquefaction reaction. A high-temperature gas-removal system is employed to provide for optimum phase separation and nitrogen removal.

Status   Laboratory testing of the process at the Montana State University is still in the early stages of development. Current studies are directed towards ascertaining optimum process conditions including the temperature and pressure of reaction, coal-to-salt-weight ratio, and zinc chloride to potassium chloride ratio.

A commercial process based on Molten-Salt Catalysis must provide a reliable method for regeneration of the catalyst. This will entail removal of sulphur, carbon and nitrogen impurities to yield a relatively pure potassium chloride or zinc chloride melt for recycling to the hydro-cracking step.

A molten-zinc chloride catalysed coal-extract hydrotreating pilot plant is included in the C.S.F. pilot plant of Consolidation Coal Company at Cresap, W. Va. This plant is to provide data for the commercial application of the process.

PROJECT LIGNITE
(also known as "Solvent-Refined Lignite (S.R.L.))

Process Developer: University of North Dakota, Dept. of Chemical Engineering, Grand Forks, N.D.

Sponsors: Great Northern Railway Company, St. Paul, Minnesota.
Office of Coal Research.

Description: The process can best be described as a lignite refinery concept in which liquefaction, gasification, hydrogenation, drying, and carbonisation processes are evaluated for possible integration or combination to produce pipeline gas, synthetic crude oil, low-sulphur, low-ash solid fuel, industrial coke and various by-products. The probable route involved will incorporate solution-hydrogenation of lignite using an organic solvent and a synthesis gas. Process solvent is generated from the lignite and recycled in continuous operation. A further step, catalytic hydrogenation of the solvent-refined lignite to produce liquid hydrocarbons, will be incorporated in the process.

Operating Conditions: Few details are available on operating conditions, however they should be similar to those for PAMCO's S.R.C. process.

Status: In the 1960's, Great Northern Railway Company sponsored preliminary investigation of the solvent-refined coal process by the University of North Dakota with the cooperation of the Spencer Chemical Company (now Pittsburgh and Midway Coal Mining Co). The objectives of this early work were to determine favourable conditions for the solvent refining of lignite by the Pott–Broche process, a German process on which patents date back to 1927 and upon which modern S.R.C. processes are based, and to investigate some of the uses for the de-ashed product.

Batch autoclave studies in a one gallon capacity stirred reactor were carried out to determine:

1. Whether lignite could be successfully dissolved in solvents that might be practical for commercial development.

2. Whether the solvents would function after successive cycles of recovery and reuse.

3. The extent of ash removal from the filtrate and the nature and quantity of mineral matter passing through the filter.

During these autoclave studies optimum operating conditions and solvent type were established.

The development of the lignite refinery concept has been sponsored by O.C.R. since 28th March, 1972 under a 5 year, U.S. $3,356,000 contract. The basic concept of the research effort is to indentify specific steps in the overall processes that need development beyond the laboratory stage, to conduct bench-scale research for obtaining information necessary to establish continuous operation, and to design, construct and operate process development units where necessary to obtain data for a pilot plant for the integrated process.

Economic analyses of projected processes for the conceptual refinery are being done at various steps to ensure that the overall integration will be both economically and technically feasible.

The first process chosen for study in a P.D.U. was that of solution-hydrogenation with synthesis gas. To this end, a 50 pound per hour P.D.U. to test this process step has recently been completed at Grand Forks. This P.D.U.

25

produces solvent-refined lignite from lignite slurried in a heavy aromatic solvent and synthesis gas (CO and H_2).

A second stage, not to be operated for perhaps another year, will upgrade S.R.L. to hydrocarbon liquids by fixed-bed catalytic hydrogenation.

The Ralph M. Parsons Company assisted in specifying and selecting instrumentation and control equipment for the P.D.U's.

PROJECT SEACOAL

This is a project undertaken by the U.S. Department of the Navy in which extensive evaluations are being made of the utilisation of synthetic fuels derived from coal substitutes for petroleum products for propulsion of its ships and aircraft.

In Project Seacoal I, evaluation of syncrude from the COED pilot plant at Princeton, N.J. has been undertaken. A U.S. Navy destroyer, the U.S.S. Johnston, has been fueled by COED syncrude in successful test runs.

The overall project, of an estimated three-years duration, will test synthetic fuels in pumps, naval boilers and gas turbines. The awareness of the strategic advantage of the utilisation of reserves of coal-derived fuels by the Navy has been responsible for the initiation of the project.

PROJECT SEACOKE

Process Developer: Arco Chemical Co. Philadelphia, Pa., a division of Atlantic Richfield Co.

Sponsor: Office of Coal Research.

Description: Coal, usually blended with petroleum residuum as part of the feedstock, is subjected to pyrolysis in multi-stage fluidised beds to produce char, liquid products and gas. After separation of the phases, the liquid products of pyrolysis are further catalytically hydro-treated to produce gasoline. The gas can be utilised in the process or processed for sale and the char would ideally be used for power eneration "on site".

No details are available on operating conditions.

Status: This project was operational under O.C.R. sponsorship during the 1960's. In January, 1970, the final report on the project was issued by O.C.R. This followed a report from Blaw-Knox Chemical Plants Division (Dravo Corp.) under a O.C.R. sub-contract, on a commercial plant design and engineering estimate of a Project Seacoke facility for the utilisation of coal and residual fuel oil to produce char, syncrude and fuel gas. A 10,000 barrel per day plant for location in the St. Louis area, was designed. This Blaw-Knox project successfully fulfilled the main requirement of Atlantic Richfield's prime contract with O.C.R. Support for the project was stopped at this stage and O.C.R. attention on pyrolysis transferred to the C.O.E.D. project which was considered to be less complex than Project Seacoke.

Note: Since 1968, Atlantic Richfield has been a co-sponsor of the H-COAL process. Refer to "H-COAL" in this section of the report for additional information.)

SOLVENT EXTRACTION–C.R.E.

Process Developer: Coal Research Establishment, Stoke Orchard, United Kingdom.

Sponsor: National Coal Board of the United Kingdom.

Description: Coal is digested in a hydrogen-donor solvent (e.g. anthracene oil) to yield a slurry from which mineral matter and unreacted coal is separated by filtration, leaving a coal solution suitable for the production of liquid fuels by further catalytic hydrogenation, or for the production of electrode coke or carbon fibre.

Status: In an experimental plant at Stoke Orchard coal can be continuously digested at a rate of 15 kg/hour and the slurry product filtered and either converted into solvent refined coal (S.R.C.) by evaporation of the solvent, or coked. This plant is used to confirm the results of laboratory-scale experiments, to develop continuous methods for solvent extraction, to evaluate the production of electrode coke, and to prepare extract for the preparation of carbon fibre.

The research program to date has shown that a calcined coke of premium electrode-making quality can be produced from coal extract, that carbon fibre of commercial quality can be produced from solid S.R.C., and that by catalytically hydrogenating the coal extract a range of liquid hydrocarbon fuels can be prepared.

(See also "GAS EXTRACTION" in this section of the report.)

SOLVENT EXTRACTION–U.O.P.

Process Developer: Universal Oil Products Co.

Description: Coal and solvent are introduced into the reaction vessel at sufficient velocity to cause degradation of the coal by impact against a fixed plate located in the reactor. The slurry residence time in this reactor is sufficient to cause solution of most of the coal in the solvent under operating conditions of 700–1,000°F. and 500 p.s.i. Sufficient hydrogen is passed upwards through the slurry in the reactor to cause hydrogenation of the coal constituents and to maintain the hydrogen content of the solvent at about the same level of the fresh solvent. A conventional hydrogenation catalyst may be added to the reactor to assist the hydrogenation reactions.

The liquid product slurry is sent to a separator where gas is separated for recycling to the dissolution reactor. The liquid coal extract (containing insoluble residues) is passed to an extraction zone where the solid residue is removed. The liquid product stream then undergoes fractionation to produce gas, recycle solvent, light hydrocarbons and a heavy hydrocarbons bottom product. The liquid hydrocarbons are subjected to catalytic hydrotreating to produce synthetic crude oil.

Several variations on this basic route are available including placing the filtration step directly after the digester, and the utilisation of two-stage or three stage solvent extraction.

Operation Conditions:

Reactor	Temp. °F	Press.p.s.i.	Reactants	Products
Solvent Extraction	700–1,000	500	Coal-solvent -Hydrogen	Syncrude and Gas

Status:	Universal Oil Products is the holder of many basic patents relating to the solvent extraction of coal under hydrogen pressure, either with or without an hydrogenation catalyst.

The process has been tested for over two years in a small-scale pilot plant at U.O.P.'s research centre in Des Plaines, Ill.

Plans are under consideration for the construction of a demonstration plant utilising information gained from the operation of the pilot plant which has been processing Illinois coal containing about 3.8% sulphur to produce about 4 barrels of synthetic crude oil per ton of D.A.F. coal. The synthetic crude is suitable for further refining by conventional technology into gasoline, jet fuel and diesel oil.

(Refer to the listing of patents in this report for details on the broad base U.O.P. has in the solvent extraction of coal.)

SOLVOLYSIS

Process Developer: The Japanese Agecny of Industry and Technology, Tokyo, Japan.

Description: Pulverised coal is mixed with an asphalt solvent and introduced into a reactor where the mixture is heated to 400–450°F. Crystalline micro-beads are formed which absorb the ash content of the coal. These micro-beads are separated from the product stream and the liquid product can be hydrogenated to produce a wide range of synthetic fuels.

Status: The Agency of Industry and Technology has been testing the process in a 2 litre capacity reactor.

S.R.C.–C.F.C.

Process Developers: Old Ben Coal Company, a subsidiary of Standard Oil of Ohio.

Description: Few details are available on this process. Elements of the CO_2 ACCEPTOR, C.S.F. and Pamco S.R.C. processes are involved.

Status: Old Ben Coal Co. and Consolidation Coal Co. have formulated and proposed to the Office of Coal Research a 5 year, U.S. $83 million program to construct and operate a 900 T.P.D. demonstration plant. Their suggested program covers a wide range of liquefaction conditions ranging from non-catalytic hydrogenation to use of a donor solvent. It is believed the demonstration plant, to produce liquid and solid products from coal, will combine elements of the Ft. Lewis and Wilsonville S.R.C. pilot plants, the Rapid City CO_2 ACCEPTOR pilot plant, and the H-donor method of the Cresap C.S.F. pilot plant.

The demonstration plant will be located next to SOHIO's Toledo, Ohio refinery which offers a ready source of hydrogen, the required utilities, and existing buildings for laboratory work, maintenance, rail, coal-handling and other facilities to result in a saving of U.S. $10 million for the project. The product S.R.C. will be used as fuel for the adjoining Toledo Edison generating station.

In early 1974, Consol and Old Ben announced they would jointly advance U.S. $600,000 in needed engineering and equipment costs to keep the "Clean Fuels from Coal (CFC)" project on schedule, rather than wait for needed financial support from the U.S. Federal Government and from other interested participants.

It is believed that Old Ben Coal Co. has carried out testing of the process in a pilot plant. Consolidation Coal Co. operated the Cresap C.S.F. pilot plant from 1967 to 1970 and is currently operating the Rapid City CO_2 ACCEPTOR pilot plant.

S.R.C.–COMBUSTION-ENGINEERING

Process Developer: Combustion-Engineering, Inc., and its subsidiary, C-E Lummus Co., Engineering Development Centre, Bloomfield, N.J.

Description: No details have been released on this process however it is believed to be similar to the S.R.C. process of Pittsburgh and Midway Coal Mining Company with provision for catalytic hydrotreatment of the solvent-refined coal to produce stable liquid fuels for use by industry and electric utilities.

Status: C-E Lummus has been developing the process for several years. Sparse reports indicate that a major development has been made in a critical process step that will hasten the commercial application of the process in a plant expected to cost U.S. $300 million.

S.R.C.–MITSUI

Process Developer: Mitsui & Co., Mitsui Mining Co., Mitsui Coke Co., and Mitsui Kozan Coking Industry Ltd., members of the Mitsui Group of Companies, Tokyo, Japan.

Description: No details are at hand on this process, however it is believed that the Mitsui Group along with other SRC researchers exchange technical data and it is assumed that elements of the PAMCO S.R.C. process are incorporated in the Mitsui S.R.C. route.

Status: The Mitsui Group have conducted research and development work for many years on coal conversion processes. Early studies involved the operation of a 2 T.P.D. pilot plant in Fusan, Manchuria in 1928, based on the Bergius process. During World War II, Fischer-Tropsch plants were operated to produce synthetic fuels from coal.

Since the War, Mitsui has conducted bench-scale testing of various liquefaction processes. In 1970, Mitsui began research on the S.R.C. process in bench-scale units, and since September, 1973, a continuous 12 pound per hour bench-scale unit has been in operation. A 5 T.P.D. pilot plant to conduct further S.R.C. studies was commissioned in Omuta, Kyushu, in October, 1974, with operation expected in 1975. Plans are in hand to construct a plant to produce S.R.C. on a semi-commercial scale in Miike, Kyushu, this plant to be operational in 1978 with a capacity of 1,500-2,000 T.P.D. of Miike coal.

In November, 1974, Mitsui announced that samples of coal supplied by Millmerran Coal Pty. Ltd., of Brisbane, Queensland, Australia, from areas held by the company on the Darling Downs in south-east Queensland had displayed high reactivity in bench-scale tests in Japan and was considered one of the premium coals for conversion by the Mitsui process. For some time, Mitsui has been evaluating a wide range of foreign coals for suitability as S.R.C. process feedstock and the Millmerran Coal sample tested had shown that this coal was among the most suitable for conversion.

As a result of the promising results, Mitsui and Millmerran Coal announced that they would undertake a joint-venture project for the further evaluation of the Darling Downs coal (which is a very-high volatile bituminous coal) including the continuation of exploration and detailed testing and feasibility studies on the construction and operation of a commercial S.R.C. plant of 10,000 T.P.D. coal feed based on the Millmerran Coal company's reserves which stand at about 450 million tons, largely recoverable by open-cut mining. Mitsui expects this plant to be operational in late 1980. It is envisaged that this commercial plant would also incorporate a separate gasification plant to gasify S.R.C. plant carbonaceous residue together with around 1,000,000 ton per year of coal feedstock to produce significant quantities of low-, med-, and high- BTU gases. The Millmerran S.R.C. plant would either produce a solid S.R.C. product containing less than 0.1% ash and with a calorific value of about 16,000 BTU per pound, or by, utilising a modified S.R.C. process being developed by Mitsui, heavy and light oils, S.R.C., gas and chemical by-products. Cost of the S.R.C. plant is estimated at A $300 million.

(See also "S.R.C.–PAMCO").

S.R.C.–PAMCO

Process Developer: The Pittsburgh and Midway Coal Mining Company, Ft. Lewis, Washington, a subsidiary of Gulf Oil Corporation.

Sponsor: Office of Coal Research.

Description: Raw coal is pulverised and mixed with a coal-derived solvent boiling in the general range 550-800°F. Hydrogen, or a hydrogen-rich synthesis gas, is added to the coal-solvent slurry and passed through a preheater to a dissolver vessel. The dissolver is operated at 815°F and 1,000 psi with an excess of hydrogen, and under these conditions approximately 90% of the D.A.F. coal is dissolved. The actual degree of dissolution of the coal depends on the "reactivity" of the particular coal feed. In addition to solution of the coal, several other major types of reactions occur. These are: (1) depolymerisation of the coal, necessarily accompanied by hydrogenation of the coal; (2) hydro-cracking of the solvent to lower molecular weight hydrocarbons ranging from light oil to methane; (3) removal of organic sulphur by hydrogenation of the sulphur to hydrogen sulphide.

The product stream from the dissolution-hydrogenation step consists of coal solution, unreacted coal (inerts), undissolved mineral matter (ash), light hydrocarbon gases (methane-rich) and excess hydrogen. The excess hydrogen and light hydrocarbon gases are separated from the product slurry. A portion of the hydrogen stream is recycled to the dissolution reactor and the remaining hydrogen, together with the light hydrocarbon gases are further processed for utilisation as plant fuel or for sale as "town" gas (600 BTU/SCF) or upgraded by methanation to pipeline-quality gas (1,000 BTU/SCF).

The product slurry is pumped to the filtration section where the undissolved coal solids are separated from the coal solution. The filtrate is sent to a vacuum-flash distillation step for removal of the solvent for recycle to the reactor. The bottoms fraction from the vacuum-flash tower is a hot liquid with a solidification point of about 350°F. This is the major product of the process and is referred to as "Solvent-Refined Coal(S.R.C.)". This material can either be transported as a hot, molten liquid or solidified by cooling for shipment.

Alternatively, the process could be modified to produce products that are liquids at normal temperatures and pressures by subjecting the product slurry, after filtration and solvent recovery, to hydro-cracking and subsequent hydro-treatment. Hydro-cracking is employed firstly to product a lighter, hydrogen-enriched material and also effect bulk removal of hetero-atoms such as sulphur, nitrogen and oxygen which would increase costs in refinery operations. A tar by-product is obtained from this step and this tar may be mixed with the filter cake from the filtration step (which contains about 50% carbonaceous matter and 50% ash) and gasified in a commercially-available gasifier to produce the hydrogen for the process. The product liquid from the hydro-cracking step is further hydro-treated to produce light refinery liquids.

Off-gases from the hydro-cracking and hydro-treatment sections are combined with off-gases from the S.R.C. process and sent to acid-gas absorption for removal of CO_2 and H_2S prior to being sent to a hydrogen-methane cryogenic separation unit. The resultant hydrogen stream is recycled to the hydroconversion section and the methane is available for sale as pipeline gas.

(Refer to C.O.G. Refinery Concept in this section of the report for a more detailed discussion on the production of light refinery liquids from S.R.C.).

Operating Conditions:

Reactor	Temp.°F	Press.p.s.i.	Reactants	Product
Dissolver	815	1,000	Coal-Solvent-H_2	Solvent-Refined Coal (S.R.C.) Gas

Status:

The early work on the solvent refining of coal in solution under hydrogen pressure was carried out by A. Pott and H. Broche in Germany in the 1920's. Various patents describing the Pott-Broche process are listed:

"Destructive Hydrogenation of Coal in Solution", A. Pott and H. Broche; British Patent 293,808; 12th July, 1927. "Hydrogenating Coal", A. Pott and H. Broche; French Patent 657, 409; 27th June, 1928.

"Extraction and Hydrogenation of Coal", A. Pott and H. Broche; U.S. Patent 2,308, 247; 12th January, 1943.

In this process, the coal is dissolved in a solvent, in a high-temperature, high pressure reactor. The product crude degrade is filtered to remove insolubles and the filtrate is subjected to distillation to recover solvent for recycle. The liquid bottom product of distillation is hydrogenated under pressure. Hydrogenated product is separated into a product stream and a sludge which is recycled to the crude degrade. Distillation of the product stream produces benzene, heating oil and liquid fuels.

The Pott-Broche process was utilised in Germany during World War II for the production of a raw material for carbon electrodes for aluminium plants.

During the 1950's, research and development work on a modification of the Pott-Broche process was carried out by Spencer Chemical Company. In 1962, the Office of Coal Research (O.C.R.) awarded a research contract to Spencer Chemical to evaluate the technical feasibility of the Solvent-Refined Coal (S.R.C.) process as it was then termed. This contract was concluded in 1965 upon the successful completion of the demonstration of the process in a 50 pound per hour continuous-flow process development unit. During the currency of this contract, Gulf Oil Corporation acquired Spencer Chemical Company and reassigned the S.R.C. project to the Research and Development Department of The Pittsburgh and Midway Coal Mining Company (PAMCO).

On 10th October, 1966, O.C.R. awarded a 9.5 year contract to PAMCO to continue research and development of the S.R.C. process. This U.S. $28,416,437 contract provides for a study of the commercial feasibility of the process through design, construction and operation of a pilot plant to process 50 ton per day of coal. The objectives of the pilot plant program are:

1. To further develop the S.R.C. process.

2. To provide design data for future commercial plants.

3. To provide bulk samples of Solvent-Refined Coal and process by-products for market-development studies.

The Stearns-Roger Corporation completed the design of the pilot plant in 1969. A shortage of funds delayed the start of construction until January, 1972 when a contract for the detailed engineering and construction of the pilot plant was awarded to Rust Engineering Co. of Birmingham, Ala. Field construction was underway in July, 1972 with completion and preliminary start-up in mid-October, 1974.

The only work remaining involves several major modifications to the pilot plant including the addition of an emergency power system, a biological waste treatment system, a fire water booster pump and a slurry recycle system. This latter system is being installed as laboratory work has shown that recycle of unfiltered process slurry to replace all or part of the solvent in the process can substantially increase the conversion of coal. This modification makes it possible to produce S.R.C. of lower sulphur content and lower viscosity than in the basic process without slurry recycle. All the equipment for the slurry recycle system has been ordered and should be installed in the pilot plant by June, 1975. The plant has operated for brief periods at 50% of the design capacity.

The 50 T.P.D. pilot plant is located at Ft. Lewis, Washington. The pilot plant includes all the major steps in the conceptual S.R.C. process except the step for the production of hydrogen (or synthesis gas) from the process residue which is under study in several O.C.R.-sponsored projects. The make-up hydrogen for the dissolving section is generated in the pilot plant by the stream-reforming of natural gas in a packaged type of plant having a capacity of 800,000 SCF per day. As well as hydrogen, the natural gas reforming unit can also produce a synthesis gas (50% H_2, 50% CO) so that tests can be run in the pilot plant to evaluate yields with both pure hydrogen and synthesis gas.

Bench-scale tests at PAMCO's Merriam, Kansas, research laboratory have suggested that with some coal types, the conversion of the coal to S.R.C. is optimised by the utilisation of synthesis gas. The use of synthesis gas instead of hydrogen would improve the economics of a commercial S.R.C. plant. These bench-scale tests at Merriam are continuing to provide back-up data for the operation of the pilot plant at Ft. Lewis.

In Wilsonville, Ala., a 6 T.P.D. pilot plant based on PAMCO's S.R.C. process has been operating for one year to provide design data for the Ft. Lewis pilot plant. This plant was procured, constructed and is operated by Catalytic, Inc., of Philadelphia, Pa., under the sponsorship of The Southern Company of Atlanta, Edison Electric Institute and the Electric Power Research Institute. On 23rd December, 1974, the Wilsonville pilot plant completed a 75-day run on Illinois No. 6 coal feed containing 3% sulphur and 10-12% ash on a dry basis. Sulphur content of the S.R.C. product was 0.6-0.9% and ash content less than 0.15%. Further long-duration runs are planned to optimise critical operations of the S.R.C. process including the utilisation of a plate filter and hydrocyclones for residue removal, and solidification of the S.R.C. by water quenching on a shaker conveyor.

The future operation of the Ft. Lewis pilot plant, as well as providing data for the design of a larger-scale demonstration plant, will provide bulk samples of S.R.C. for detailed testing, including combustion tests to be performed by the Babcock and Wilcox Company. The calorific value of solid S.R.C. is about 16,000 BTU per pound and its composition is uniform regardless of the type of coal processed. Ash contents of about 0.1% and the removal of about 90% of the original sulphur from the coal are typical of the S.R.C. product. As well as providing a low-sulphur, very-low-ash, high-calorific-value material for combustion, S.R.C. can be utilised for the production of high-grade metallurgical formed-coke from non-coking coals by the correct blending of S.R.C. with the coal, or can be used directly in a delayed or extractive coker to give volatile products and carbon.

Solid S.R.C. may also be further hydrogenated to produce a range of liquid fuels or refinery feedstocks. To oversee the commercial development of the synthetic-fuel processes, Gulf Oil Corp. has formed a new company, Gulf Energy and Minerals Co. (GEMCO), which includes the Pittsburgh and Midway Coal Mining Co. and another Gulf subsidiary, Gulf Mineral Resources.

Under various O.C.R. contracts, numerous studies are underway on various aspects of the S.R.C. process.

The Ralph M. Parsons Company have prepared a preliminary conceptual design of a C.O.G. refinery plant in which a modified S.R.C. process and the BI–GAS gasification process are combined to produce liquid fuels, refinery feedstocks, chemicals and substitute natural gas.

Chem Systems Inc., with the cooperation of PAMCO, have also prepared an economic evaluation and conceptual design of a processing complex for the conversion of coal to oil and gas in a C.O.G. refinery employing a modified S.R.C. process coupled with a coal gasification plant. The economics of several coal gasification processes for producing methane from a mixture of fresh coal feed (50%) together with the filter cake from the S.R.C. process and tar from the hydroconversion step (50%) are also under evaluation.

(Refer to the C.O.G. Refinery Concept in this section of the report for additional information on the conceptual commercial-scale S.R.C. process applications.)

Other S.R.C.-related studies forming part of the program through O.C.R. sub-contracts are:-

1. Washington State University is studying the utilisation of the mineral residue (ash) from the S.R.C. process e.g. road fill, cement, fertiliser, and as the pre-coat for the S.R.C. process rotary filter as a substitute for the more expensive diatomaceous earth.

2. Colorado School of Mines, Chemical and Petroleum-Refining Engineering Department, Golden, Colo. for a study on coal desulphurisation by the S.R.C. process under a 3-year contract funded by O.C.R. (U.S. $148,228), State of Colorado (U.S. $25,000), and Colorado School of Mines (U.S. $11,792). The contract was completed on 24th March, 1975.

3. Foster Associates, Inc., Washington, D.C., to measure the potential regional markets for coal-derived products, including S.R.C., over the period 1975 to 1985. The study is to provide guidelines to O.C.R. as to regional market opportunities for multi-product coal conversion plants (e.g. C.O.G.) which may be located in the coal fields of Appalachia, Four Coners, Fort Union and the Illinois Basin.

4. Bonner & Moore Associates, Inc., Houston, Tex., to establish the value of crude oil synthesised from coal.

5. Oklahoma State University is studying the removal of sulphur and nitrogen from coal-derived liquids.

6. Carnegie-Mellon University is conducting an environmental impact study to determine the effects of a commercial S.R.C. plant on an area encompassing the Ohio River Valley.

7. Stanford Research Institute, Menlo Park, Calif., to determine the economic feasibility of a coal conversion plant, either H–COAL or S.R.C., in

South-western Alaska utilising the Beluga coal reserves. Under a 16-month U.S. $201,023 contract funded by O.C.R. to U.S. $153,023 and jointly by Nissho-Iwai American (a subsidiary of Nissho-Iwai of Japan) andPlacer Amex to U.S. $48,000. This study will include an evaluation of the potential markets for the products in Japan and on the west coast of the U.S.A.

8. Ralph M. Parsons Company for technical evaluation of the performance of the pilot plants and other experimental operations, preliminary design of a commercial-scale S.R.C. plant and an economic evaluation of this commercial plant.

9. Dow Chemical Company, to evaluate which of three coal-conversion processes, C.O.E.D., H–COAL or S.R.C., could most effectively be used in a commercial plant.

Related Developments:

A. In a significant recent development, Wheelabrator-Frye, Inc. has announced that it will finance, own and operate a 1,000 ton per day demonstration plant based on PAMCO's S.R.C. process at a site yet to be selected. Southern Services, Inc., a subsidiary of the Southern Company which is the sponsor of the Wilsonville, Ala. 6 T.P.D. S.R.C. pilot plant, and Wheelabrator-Frye have contracted to utilise PAMCO's S.R.C. process under an arrangement with Gulf Oil Corp. The project will commence with a feasibility study and the construction of the 1,000 T.P.D. demonstration plant estimated to cost U.S. $70-100 million. Should this demonstration plant operate successfully, it will be expanded at a cost of U.S. $350 million over 3 years to produce 10,000 T.P.D. of S.R.C. The S.R.C. would be consumed in electric generating stations in The Southern Company's system.

B. In another significant development, the Mitsui Group recently announced a joint-venture which it is hoped will expedite the establishment of commercial plants probably based on joint American-Japanese techniques, modifications and experience. Mitsui Co. of Japan intends to joint-venture with a Brisbane, Queensland-based company, Millmerran Coal Pty. Ltd., in a feasibility study for an S.R.C. plant on the Darling Downs in southern Queensland, Australia, which will utilise the extensive reserves of very-high-volatile bituminous coal held by Millmerran Coal. The anticipated scale of the plant for Millmerran would process 10,000 T.P.D. of feedstock coal, or 3 million ton per year, with 1 million ton per year being fed to a separate gasifier for the production of significant quantities of substitute natural gas, hydrogen or synthesis gas for ammonia or methanol production. Estimated cost of the plant is about A$300 million. The plant is scheduled for operation during 1981. Successful operation of the first commercial S.R.C. plant could pave the way for further and various coal-conversion plants on the Darling Downs to process the large reserves of coal in the Millmerran Coal Company's areas with the ultimate aim of self-sufficiency for the State of Queensland in oil, gas, Solvent Refined Coal and a wide range of chemical by-products that would provide the basis for a petro-chemical industry. Reserves of the highly reactive bituminous coal held by Millmerran Coal are believed to be in excess of 450 million tons, mostly recoverable by open-cut mining, with excellent prospects for the delineation of larger reserves by exploration in the future.

(Refer to "S.R.C.–Mitsui" in this section of the report for further details.)

SYNTHOIL

Process Developer: U.S. Bureau of Mines, Pittsburgh Energy Research Centre, Pittsburgh, Pa.

Sponsor:	U.S. Bureau of Mines, Office of Coal Research and Bethlehem Steel Corp.

Description: Pulverised, dried coal is slurried in a recycled portion of its own product oil. The slurry is pumped into a catalytic, fixed-bed reactor with hydrogen at high velocity to create turbulent-flow conditions. The reactor is filled with immobilised catalyst pellets composed of cobalt molybdate on silica-activated alumina. The combined effect of the hydrogen, turbulent-flow conditions and catalytic action is to liquefy and desulphurise the coal. Operating conditions within the reactor are 850^OF and 2,000-4,000 psi, and a residence time of 15 minutes allows for high yields and high throughput.

The product passes through a high-pressure receiver where gas is separated and recycled after ammonia and hydrogen sulphide removal. The raw oil is centrifuged to remove ash and unreacted coal, providing a low-sulphur, low ash fuel oil.

Part of the product stream is recycled to slurry preparation. The remaining portion of the product oil can be used as a fuel oil or refined further to gasoline or diesel fuel if desired.

Operating Conditions:

Reactor	Temp^{O}F	Press. p.s.i.	Reactants	Product
Catalytic, fixed bed	850	2,000-4,000	Coal-H_2-oil	Fuel oil, Gas

Status: The process has been tested in a 5 lb./hour bench-scale unit and a 0.5 T.P.D. pilot plant for a number of years at the Pittsburgh Energy Research Centre.

Foster-Wheeler Corp., Livingston, N.J., has recently been awarded, by the Bureau of Mines, a U.S. $6,928,416 contract for the design and engineering services for an 10 T.P.D. pilot plant to be constructed in Bruceton, Pa. The contract requires Foster-Wheeler to procure U.S. $4.5 million worth of equipment for the pilot plant which, by integrating design and construction phases, is designed to cut five months from the overall time required to bring the pilot plant into operation. The U.S.B.M. hopes in this way to begin start-up runs in 1976.

This 10 T.P.D. plant will be designed to produce 1,000 gallons per day of low-sulphur fuel oil and the operation of the plant will provide design data for a commercial-scale plant which is expected to convert 20,000 to 30,000 T.P.D. of coal into 3 to 4 million gallons of synthetic crude oil.

The overall cost of the new pilot plant is estimated at U.S. $14 million with the Bureau of Mines contributing U.S. $13 million and O.C.R. contributing U.S. $1 million.

Under a recently-announced cost-sharing agreement between Bethlehem Steel Corp. and the Bureau of Mines, Bethlehem Steel will operate the 8 T.P.D. pilot plant at Bruceton on behalf of the Bureau for about two years after startup in 1976, contributing U.S. $1.1 million towards the plant's operating expenses and U.S. $500,000 for research.

The E.R.D.A. (formerly O.C.R.) may soon start preliminary design work for a 700 ton per day SYNTHOIL pilot plant for construction in 1977 to follow the 10 ton per day pilot plant.

TOSCOAL

Process Developer: The Oil Shale Corporation (Tosco), Rocky Flats Research Centre, Golden, Colorado.

Sponsor: The Oil Shale Corporation.

Description:

Preheated coal feed from a dilute-phase fluidised bed is fed to a pyrolysis vessel where it is contacted with heated ceramic balls. The char produced in this reactor is screened, cooled and stockpiled. The cooled ceramic balls from the reactor pass over the screen and are recycled to the ceramic balls heater via an elevator. The vaporised products of the pyrolysis step are condensed and fractionated. Offgas (600 BTU/SCF) from the condensor can be utilised as a fuel gas for the ceramic pellet heater or processed for sale.

Operating Conditions:

Reactor	Temp.°F	Pressure	Reactants	Products
Pyrolyser	800-1,000	Atmospheric	Coal-heat	Char, Gas, Oil

Status:

Tosco has investigated this process for the low-temperature carbonisation of coal since 1970. This TOSCOAL process and the TOSCO II process for oil shale retorting are quite similar and can be represented by the same flow sheet. The TOSCO II process has completed final field testing in a 1,000 T.P.D. semi-works retort located at Parachute Creek, near Grand Valley, Colo. Construction engineering is underway for a 66,000 T.P.D. commercial plant.

Drawing upon the oil shale retorting technology of the TOSCO II process, Tosco has tested the TOSCOAL process in a 25 T.P.D. pilot plant located at the Rocky Flats Research Centre near Golden, Colo.

Yields from sub-bituminous coals tested in this pilot plant range from 0.3 to 0.5 barrel of oil per ton of raw coal, with the general range of heating values for product char representing about a 50 per cent increase over the heating value of the corresponding raw coal.

It is expected that development of the TOSCOAL process will parallel development of the TOSCO II oil shale process in the future.

UNDERGROUND LIQUEFACTION

Process Developers:

1. The Texas Company.

2. West Virginia University, School of Mines, Morgantown, W. Va.

Description:

Coal dissolution is begun "in situ" and completed at the surface. A hydrogen-donor solvent (e.g. anthracene oil) is introduced into a thick coal seam, via a well, under pressure to reduce solvent vaporisation and limit the reaction zone to the basal section of the seam. Coal solvent invades joint and bedding plants where imbibation and swelling by transfer of hydrogen from the solvent to the cohesive structure between micelles or microplatelets of coal occurs, weakening this structuring and producing coal fragments. After coal has reacted and fragments have fallen from the face of the seam, recovery of the slurry is intiated by means of introducing more solvent under turbulent-flow conditions to carry the fragments to the surface via another well drilled into the seam.

The resultant slurry would be processed at the surface e.g. further catalytically hydrogenated, to produce the desired liquid products.

Status:

The West Virginia University has carried out autoclave studies on various ranks of coal to determine the best solvent for the process, which proved to be anthracene oil. Sub-bituminous coal gave the best results at it was generally found to react, crack, crumble and dissolve to a higher degree than coals of other ranks.

The Texas Company is the holder of U.S. Patent 2,595,979 of 6th May, 1952, which describes an underground liquefaction process similar to that outlined above.

36

UNION CARBIDE

At Charleston, W. Va., Union Carbide operated a pilot plant during the late 1940's to investigate the production of chemical feedstocks from coal by direct catalytic hydrogenation. Results were not encouraging at that time, and problems were experienced in separating suitable chemical feedstocks from the hydrogenated product.

PATENTED LIQUEFACTION PROCESSES NOT AT PRESENT IN A SIGNIFICANT STAGE OF DEVELOPMENT

PATENT HOLDER	U.S. PATENT NUMBER	DATE	PATENT DESCRIPTION
Bennett Engineering Co.	3,576,734	27. 4.1971	Syncrude Production by Low Temperature Carbonization of Tar.
Compagnie Francaise des Essences Synthetiques S.A.	2,707,163	26. 4.1955	Solvent Extraction and Fractional Distillation
Fossil Fuels Inc.	3,030,297	17. 4.1962	Rapid Coal Hydrogenation.
Great Lakes Carbon Corp.	3,379,638	23. 4.1968	Ash-Free Anode Carbon Production (jointly with The Lummus Company)
Huntington Chemicals Corp.	3,107,985	22.10.1963	Continuous Distillation of Coal followed by Autogenous Hydrogenation of Volatiles.
Pyrochem Corp.	3,244,615	5. 4.1966	Catalytic Contact of Destructively Distilled Coal Volatiles.
Pyrochem Corp.	3,247,092	19. 4.1966	Incomplete Coal Liquefaction by Quadri-Phase, Low-Pressure Hydrogenation.
Texaco Development Corp.	2,572,061	23.10.1951	Hydrogenation of Coal in a Liquid Phase.
Texaco Development Corp.	2,753,296	3. 7.1956	Hydrogenation of Pulverised Coal in a Liquid Phase
Texas Company, The	2,658,861	10.11.1953	Hydrogenation of Coal.
Texas Company, The	2,664,390	29.12.1953	Coal Carbonisation and Recovery of Volatiles.
Texas Company, The	2,681,300	15. 6.1954	Coal Extraction Utilising Hydrogenated Thianaphthenes
Universal Oil Products Co.	3,503,864	31. 3.1970	Coal-Solvent Extraction Utilising Hydrogen Sulphide.
Universal Oil Products Co.	3,503,865	31. 3.1970	Microwave Coal Liquefaction.

Note: For a more complete listing of major U.S. Patents relating to Coal Conversion refer to APPENDIX II.

SECTION II

High BTU
Coal Gasification Processes

AGGLOMERATING ASH

Process Patent Holder: Union Carbide Corporation

Developer: Battelle Memorial Institute, Columbus Laboratories, Columbus, Ohio. (Initial development by Union Carbide Corp.).

Sponsor: Office of Coal Research and American Gas Association

Engineering Contractor: Chemical Construction Corporation (Chemico).

Note 1: Chemico and Union Carbide have formed a joint-venture company, Coalcon, to develop this technology commercially.

Note 2: Chemico is a subsidiary of Aerojet General Corp. itself a subsidiary of General Tire & Rubber Co.

Description: Coal, pulverised to -35 mesh, is injected near the base of the gasifier into a fluidised bed of hot ash agglomerates, then flows through the bed where the coal-steam gasification occurs. The char end product concentrates at the top of the bed where additional gasification occurs. Hot ash agglomerates from the combustor at 2,000-2,100°F enter the gasifier at a point below the char level and descend through the reaction and preheat zones where they are cooled to 1,000°F before entering the stripping zone. Here any entrained coal is removed, and stripped agglomerates are collected in a surge pot and pumped into the combustor. Char from the top of the gasifier is withdrawn continuously and fired in the agllomerating bed combustor with air where the agglomerates are reheated to 2,100°F for recycling to the gasifier.

Hot flue gases from the combustor are processed to recover heat, remove SO_2, and recover compression energy in an expander. Raw gas from the gasifier at 1800°F is processed to recover heat and remove particulates, ammonia, and sulphur compounds. The resultant clean gas with a heating value of about 300 BTU/SCF, can be processed further by shift and methanation to pipeline quality (950 BTU/SCF) or used directly to fuel turbines and boilers for electric power generation.

All types of coals can be processed without pretreatment, including highly-caking coals.

Operating Conditions:

Fluidized-Bed	Temp. °F	Press p.s.i.g.	Reactants	Product
Combustor	2,000-2,100	100	Char-air	Hot Char
Gasifier	1,800	100	Coal-steam	300 Btu Gas

Status: Components of the process were tested in the 1960's by Union Carbide and Battelle. In February, 1973, O.C.R. and A.G.A. awarded a $4.1 million contract to Battelle to develop this process over 30 months. Chemico began as a sub-contractor in February 1973 to engineer a process development unit (P.D.U.), construction beginning in June, 1974. This unit (to operate as a prototype of a commercial plant) will produce 800,000 s.c.f./day of synthesis gas. No shift-conversion, purification or methanation steps are involved however these can be added in a commercial plant. The P.D.U., located at West Jefferson, Ohio, will process 25 T.P.D. of coal when fully operational in mid 1975.

Note: This process is similar to the I.C.I. moving burden process.

AIR PRODUCTS RECYCLE PROCESS

Process Developer: Air Products and Chemicals, Inc., Allentown, Pa.

Description:	This process produces a high-BTU pipeline gas without the necessity for shift conversion and methanation steps by separating methane from the raw gas produced in the gasifier at low temperatures and recycling the mixture of hydrogen and carbon monoxide to the gasifier to produce additional methane and improve the heat input to the reactor. The offgas containing a mixture of carbon dioxide, carbon monoxide, methane, hydrogen sulphide and hydrogen, is passed through a char-removal step which recycles the char to the first stage of the gasifier. After acid-gas removal, which eliminates CO_2, H_2S and water, the synthesis gas stream is cryogenically separated to yield a methane product stream and the CO and H_2 stream is heated and recycled to the second stage of the gasifier where it reacts with coal, steam and synthesis gas from the first stage of the gasifier to produce additional methane and synthesis gas. This process reduces coal and oxygen requirements while requiring increased amounts of steam when compared to those processes utilising shift and methanation of synthesis gas. A water-gas shift converter can be added to the process to facilitate cryogenic separation.
Status:	Air Products and Chemicals, Inc. is the holder of U.S. Patent 3,779,725 pertaining to this process. No details are available on development work for the process. This process can be applied to other gasification processes for the production of pipeline gas such as Lurgi, BI-GAS, CO_2 ACCEPTOR, MOLTEN SALT, and the HYGAS project's Steam-Iron and Electrothermal processes.

ATGAS—PATGAS

Process Developer:	Applied Technology Corporation, Pittsburgh, Pa.(a subsidiary of International Systems & Controls Corp.)
Sponsor:	American Gas Association, Office of Coal Research and the Environmental Protection Agency.
Description:	Coal, crushed and dried, is injected into a molten iron bath through steam lances. Oxygen is introduced through lances located at the iron-bath surface. Coal dissolves in the molten iron where the volatiles crack and are converted into carbon monoxide and hydrogen. The fixed carbon reacts with oxygen and steam, producing additional carbon monoxide and hydrogen. Caking, high-ash and high-sulphur coals can be utilized.
	Sulphur of the coal migrates to a lime slag floating on the molten iron and forms calcium sulphide. The slag, containing ash and sulphur, is continuously withdrawn from the gasifier and desulphurized with steam to yield elemental sulphur and desulphurized slag. The raw gas from the gasifier can be used as an intermediate BTU fuel gas (315 BTU/SCF) or as a synthesis gas to produce other organic compounds (PATGAS PROCESS). In the ATGAS PROCESS, the medium BTU off-gas from the gasifier is subjected to shift conversion, purification, methanation and compression to produce an S.N.G. product (940 BTU/SCF). All types of coal can be gasified in this process.

Operating Conditions:

Process	Temp. $^{\circ}$F.	Press. p.s.i.	Reactants	Product
PATGAS	2,500	50.	Coal, Oxygen, Limestone Steam	315 BTU/SCF 66% CO, 34% H_2
ATGAS	2,500	50.	Coal, Oxygen, Limestone Steam plus Methanation	940 BTU/SCF 93% CH_4 1.1% H_2

Status:	The process has been under investigation in the laboratories since 1967. To date, the process has been demonstrated in short duration runs (30 — 40 minutes) in a 2 feet internal diameter gasifier. Plans for further development involve a larger gasifier to demonstrate long duration operation. Most of the

technology pertinent to the process already exists as discrete commercial steps in the iron and steel industry. However, the combination of these steps remains to be demonstrated on a large scale. The U.S. Environmental Protection Agency is evaluating the process for utilization of high sulphur coals within E.P.A. pollution standards.

BIANCHI

Description:

Pulverised coal entrained in a steam-oxygen stream is injected tangentially into the centre of a vortex chamber operating at 150-350 p.s.i. and at temperatures below 1,700 F. The ash is entrained in the product gas stream which after ash removal in dust cyclones has a calorific value of 440 BTU/scf which is suitable for catalytic methanation to pipeline quality gas.

Status:

A pilot plant was built in France to evaluate the production of pipeline gas from lignite. No details are available on the current status of the process.

BI-GAS

Process Developer:

Bituminous Coal Research, Inc., Pittsburgh, Pa.

Sponsor:

Office of Coal Research and American Gas Association.

Sub-Contractor:

Stearns-Roger Corp.

Additional Engineering:

Blaw-Knox Chemical Plants Division (Dravo Corp.) and The Babcock and Wilcox Company.

Description:

The gasifier consists of two stages, an upper entrained – flow vessel operated at 1400°-1700^OF and a vortex – flow lower stage operated at 2800^OF. Injected with steam into the bottom of the upper gasifier section, pulverized coal volatilizes as it mixes with synthesis gas moving upward from the lower section and undergoes partial methanization at 1400 to 1700^OF. The product gas, mixed with unreacted char, is then fed into the bottom gasifier section where it gasifies with oxygen and steam at 2700° to 3000^OF, leaving a slag. This slag, water-quenched to granular form, is dropped from the vessel by a pressure lock. The gas passes upward through the top stage, to a cyclone and shift conversion, purification and catalytic methanation. All types of coal can be used.

Operating Conditions:

Gasifier	Type	Temp. ^OF	Press p.s.i.	Reactants	Product
Stage 2 (upper)	Entrained-flow	1400-1700	1000-1500	Coal-Steam	378 BTU/SCF
Stage 1 (lower)	Vortex-flow	2800	1,000-1500	Char-O_2-Steam	(H.H.V. of Raw Gas)

Status:

Development work has proceeded from batch autoclave studies, through continuous-flow experiments in a 5 lb./hour externally-heated reactor, to operation of a 100 lb./hour internally-fired process and equipment development unit (PEDU). With completion of the PEDU test program in late 1971 and following an O.C.R./AGA 6 year $24.8 million contract, a fully-integrated 120 T.P.D. pilot plant was designed by Stearns-Roger Corp., construction beginning on 27/7/1972 at Homer City, Pa. The pilot plant should be completed in mid-75. The final objective is to provide sufficient design data for construction of a commercial plant. A fluidized-bed catalytic methanation system was installed in the pilot plant following testing in June 1973. The P.E.D.U. test methanator is capable of handling up to 6,000 cubic feet per hour of 378 BTU/SCF synthesis gas at 1650^{0}F and 1050 p.s.i.g. The scaled-up methanator for the pilot plant will treat up to 300,000 cubic feet per hour of synthesis gas from the BI-GAS gasifier.

The commercial concept derived from the BI-GAS pilot-plant program envisages a $300 million plant processing 5 million tons per year of coal to produce 250 million standard cubic feet per day of pipeline gas. Estimated cost of the gas is $1.10 per 1,000 cubic feet. Water requirement is estimated at 15,000 acre feet per year.

Note: 1. Bituminous Coal Research, Inc. is the research affiliate of the National Coal Association of the U.S.A.

2. Refer to "Multiple Fluidized Bed" in the Low & Med – BTU section of this report for operation with air.

3. The BI-GAS process has been patented by the Dept. of the Interior.

CHEVRON GASIFICATION

Process Developer: Chevron Research Company, a subsidiary of Chevron Oil Corp.

Description: The process produces high – BTU gas from a wide range of organic feeds such as lignite, organic waste materials, wood and essentially any organic material containing some hydrogen and at least 10 wt. % oxygen with 25 wt. % oxygen preferred. The organic material is reacted with steam at 50–800 p.s.i. (preferably 300–800 p.s.i.) and 1200–1400^{0}F in the presence of an alkali-metal catalyst (e.g. K_2CO_3). The high-BTU gas is produced under these conditions by the catalytic steam-reforming of the products of degradation of the feed.

Operating Conditions:

Reactor	Temp^{0}F	Pressure psi	Reactants	Product
Catalytic	1200–1400	300–800	Organic Feed-Steam	High-BTU Gas

Status: U.S. Patents 3,775,072 and 3,759,677 described by R.J. White, have been assigned to Chevron Research Company. Details on development of the process are not available.

CO$_2$ ACCEPTOR

Process Developer: Consolidation Coal Company (Consol), Pittsburgh, Pa., a subsidiary of Continental Oil Co.

Sponsors: Office of Coal Research and American Gas Association.

Sub-Contractor: Stearns-Roger Corp. Denver, Colo.

Engineering: Blaw-Knox Division, Dravo Corporation, Pittsburgh, Pa.

Description: Coal is ground, dried and fed into a fluidised-bed gasifier where, under pressure of 150 to 300 p.s.i., it is heated in the presence of steam to a temperature of 1500^{0}F. Dolomite, preheated to 1900^{0}F in a regenerator vessel and calcined, is fed into the top of the gasifier. The high temperature calcined particles filter down through the gasifier, furnishing heat by a unique chemical reaction in which the calcined dolomite absorbs carbon dioxide exothermically. The spent dolomite and carbon residue are circulated to the regenerator where the dolomite is re-calcined using heat from the burning of the char. The gases released by heat and chemical reaction between steam and coal in the gasifier contain all the ingredients needed for subsequent methanation to pipeline gas.

Operating Conditions:

Fluidised-Bed	Temp $^{\circ}$F	Press p.s.i.	Reactants	Product
Gasifier	1,500	150–300	Coal-steam-CaO	Raw gas has H.H.V. of
Regenerator	1,900	150–300	Char-air-dolomite	400 BTU/c.f.

Status:

The process was developed by Consol and carried through the laboratory stage until 1964 when O.C.R. awarded a contract now, $22.6 million, to complete the bench-scale development of the process. This phase was completed successfully in 1968. Construction of a 40 T.P.D. pilot plant in Rapid City, South Dakota, was begun in January, 1970 and completed in November, 1971 by Stearns-Roger who also operate the plant. A series of integrated runs which fully demonstrated the feasibility of the process were made in 1974 culminating in a successful 10 day run converting 30T.P.D. of coal to 1.37 million SCF/day of 400 BTU gas. Most of the major mechanical problems have been solved. Lignite has been tested in long-duration runs and in future Consol will study the influence of different feedstocks, including a Montana sub-bituminous coal and at least one alternate acceptor. On 14/6/1973, an agreement was reached with Blaw-Knox for engineering and construction of a gas clean-up methanation system based on a packed-tube reactor (which has been demonstrated commercially by Conoco at Westfield, Scotland) to be added to the Rapid City pilot plant for up-grading of the raw gas.

The Rapid City pilot plant is also being considered by the Office of Coal Research as a possible site for the Liquid-Phase Methanation pilot plant, developed by Chem Systems, Inc., which is due for completion in May, 1975. This pilot plant would be used to up-grade the synthesis gas to pipeline gas.

The commercial concept for the CO_2 ACCEPTOR process envisages a $150 million plant processing 30,000 tons per day of lignite to produce 250 million cubic feet per day of pipeline gas. The pilot plant, currently processing 40 tons per day of lignite and 3 tons per day of dolomite, produces 2 million cubic feet per day of 400 BTU/SCF synthesis gas for up-grading to pipeline gas by one of the methanation steps under consideration.

Note: Consolidation Coal Company is a subsidiary of Continental Oil Company. The Rapid City pilot plant is managed and directed by Conoco Coal Development Company which is also a subsidiary of Continental Oil Company.

DEGASIFICATION

Process Developer: U.S. Bureau of Mines.

Description: Methane evolved "in situ" in coal seams is collected, stored and used as a supplement to natural gas. Yield depends on seam permeability and this may be increased by controlled fracturing and by drilling horizontal holes into the seam.

Status: The U.S. Bureau of Mines has conducted degasification tests in Washington County, Pa., Jefferson County, Ill., and Buchanan County, W. Va., since September, 1972, one billion cubic feet of methane has been drained from two experimental Bureau degasification sites. At present methane from the degasification site in West Virginia is distributed to consumers, both residential and commercial, through the existing natural-gas pipeline system.

The Bureau is interested in developing the process on a large scale, thereby avoiding the discharge to the atmosphere of vast amounts of methane released during mining operations and in their newest project, conducted at the Mining & Safety Research Centre in Bruceton, four-seam degasification is being evaluated.

Consolidated Gas Supply Corp. has undertaken the commercial application of research gained by the Bureau of Mines. This company's pipeline system was made available to the Bureau for its degasification testing of a coal seam owned by Eastern Associated Coal Corp. in West Virginia. On 28th January, 1974, methane gas from the Bureau's project began to flow into a Consolidated Gas pipeline at the rate of 600,000 SCF/day. The gas left the coal seam via a 6 ft. diameter well. The cost of this experimental project was offset by increased coal production, improved mine ventilation and the value of the methane at sale. Consolidated Gas Supply and Eastern Associated Coal have negotiated a gas-purchase contract and are planning a second degasification project using an 18 ft. diameter exit duct about two miles from the first test site. The 18 ft. duct should deliver 965,000 SCF/day of methane. Similar projects are being mooted by other gas companies. The future commercialisation of the process lies with careful integration of mining and degasification operations.

Note: Pennsylvania State University announced in April, 1975, that the Bureau of Mines and the Pennsylvania Science and Engineering Foundation are sponsoring a 5-year degasification program. The first year's operation will entail three boreholes at the U.S. Steel Corp's Cumberland mine in Green County, Pa., and two further boreholes on state land.

ELECTRIC ARC GASIFICATION

Process Developer: Columbia University, N.Y.

Sponsor: Consolidated Natural Gas Corp.

Description: The carbon in coal reacts with steam in an electric arc at about 10,000°C. Depending on reaction and subsequent quench conditions, the process can be used to produce high–BTU substitute natural gas without an additional methanation step.

Operating Conditions:

Gasifier	Temp. °C	Reactants	Product
Electric arc	10,000	Coal-steam	High B.T.U. S.N.G.

Status: As of May, 1973, the process had been tested on a batch basis at about 30 kW.

ELECTROFLUIDIC GASIFICATION

Process Developer: Dept. of Chemical Engineering and Engineering Research Institute, Iowa State University, Ames, Iowa.

Sponsor:	Office of Coal Research.
Description:	An electrofluidic reactor utilizes a fluidised-bed of conducting particles which is heated by passing an electrical current through the bed. The bed itself serves as a resistor between electrodes placed in contact with the bed. Since heat is generated directly within the bed, the device is useful for carrying out reactions which require substantial energy inputs and are favoured by high temperatures. Reacting steam and coal char in the reactor produces a wide range of hydrogen-carbon monoxide mixtures, as well as mixtures containing methane, suitable for up-grading by methanation to S.N.G.

Operating Conditions:

Gasifier	Temp ^oF	Press p.s.i.	Reactants	Product
Fluidized-Bed	1,500	1500	Char-steam	Synthesis Gas

Status:

During the course of the research sponsored by O.C.R. to June 1974 under a $426,960 contract, both a 4" diameter batch reactor and a 12" diameter continuous reactor have been successfully operated. The Institute of Gas Technology has undertaken to integrate this process with the pilot plant testing of its HYGAS process' electrogasifier.

A preliminary estimate of the cost of manufacturing S.N.G. by electrofluidic gasification of bituminous coal char with steam has indicated that this process may be competitive with other processes under development. Overall plant efficiency would be high, however a cheap source of electrical power would be necessary for the process to be economical.

EXXON GASIFICATION

Process Developer:	Esso Research and Engineering Co., Baytown, Tex. (a subsidiary of the Exxon Corporation, New York.)
Description:	Coal is reacted with steam in a fluidised-bed gasifier at 1500-1700^oF. To provide the necessary heat, a stream of circulating char is withdrawn from the gasifier and partially burned with air in a char heater to raise its temperature. The heated char is returned to the gasifier after separation from the flue gas. The product gas is a medium BTU gas suitable for methanation to S.N.G. As a high BTU gasification process Exxon's route is unique in that air is used rather than the oxygen most other processes use, thus eliminating the need of an oxygen plant. All types of coal can be gasified.

Operating Conditions:

Fluidised-bed	Temp. ^oF	Reactants	Product
Gasifier	1500–1700	Coal–steam	Med. BTU Gas suitable for methanation
Char heater	1700+	Char–air	

Status:

Exxon, and its affiliate, Carter Oil Company, have been engaged in coal gasification research since 1966 and has expended in excess of $20 million on this process. In 1969, Carter Oil became responsible for Exxon's coal and oil shale research and development. A 0.5 ton per day integrated pilot plant has been in operation at Baytown for some years. Plans for a 500 T.P.D. pilot plant at Baytown were announced in early 1974 when Carter Oil Co. awarded a $40 million contract to Arthur G. McKee & Co. for construction of the pilot plant to begin in late 1974 and completion by late 1976. Subsequently, Exxon Corp. announced it had deferred construction of the plant and would continue research on smaller-scale tests. Deferral was reputedly caused by rising costs and competition for finance from other projects.

GARRETT'S COAL GASIFICATION

Process Developers: Garrett Research & Development Co. and Island Creek Coal Co., subsidiaries of Occidental Petroleum Corp.

Sponsors: Occidental Petroleum Corp. and Colorado Interstate Gas Corp.

Description: The process is essentially similar to Garrett's Coal Pyrolysis process for the production of solid, liquid and gaseous fuels. This pyrolysis process utilises a low-temperature pyrolysis step to optimise production of liquid fuels whereas for high yields of methane-rich gas in the gasification process, the coal feed is subjected to a rapid, high-temperature pyrolysis in the reactor.

Pulverised coal is fed to the pyrolysis reactor with a recirculating stream of hot char. This hot char is from a separate, air-blown char heater which exhausts the nitrogen-rich gas produced as a flue gas. The hot char from the heater circulates to the pyrolyzer and provides the heat for the pulverized-coal pyrolysis. The pyrolysis gas is separated from entrained char in a series of cyclones and is then sent to purification, shift and methanation to upgrade the gas from its raw-state H.H.V. of 600–650 BTU/SCF to pipeline quality. The product char is a fine, highly reactive fuel suitable for combustion in a power-generation station. The calorific value of the product char can be significantly higher than that of the coal feed.

Operating Conditions: Few details are available on the operating conditions for this process. It is known to be a high-temperature, atmospheric reaction in the pyrolysis reactor, with assumed high-temperature, atmospheric, air-blown conditions in the char-heater vessel.

Status: Garrett originally concentrated R & D work on the low-temperature GRD Coal Pyrolysis process for the production of liquid fuels. This process is being tested in a 3.6 T.P.D. pilot plant in La Verne, Calif. Using this test work as a base, Garrett, in conjunction with Colorado Interstate Gas, initiated development of the coal-gasification process in 1971. The process has been successfully evaluated in a pilot plant. The results of the pilot plant testing provided data which were evaluated by Combustion Engineering subsidiary, Lummus Co., to provide a detailed design for a commercial plant. The commercial plant would process 40,000 T.P.D. of coal to yield 250 million S.C.F./day of pipeline-quality gas and sufficient char to feed a 1200 MW power plant. Estimated cost of the commercial plant is U.S. $310 million (1974) which appears to be considerably less expensive than equivalent S.N.G. plants based on existing technology. Estimated cost of pipeline gas is U.S. 94 cents per million BTU.

Garrett plans a 250 T.P.D. demonstration plant to be located near a power utility to be selected. Design and construction of the commercial plant could be commenced during operation of the demonstration plant.

Operation of the commercial-scale pyrolysis reactor has been simulated with the operation of a continuous 3 pound per hour laboratory-scale reactor which had the same configuration as the projected commercial unit except that its heat source was electrical. Results have indicated yields of pipeline-gas equivalents ranging from 4500 SCF/ton of D.A.F. coal at 1500°F to 7500 SCF/ton   (D.A.F. basis)   at 1700°F, depending on coal type. Commercial-scale operation should yield an additional 1,000 SCF/ton at 1700°F of pipeline-gas equivalent by recycling the tar produced in the pyrolysis step to the reactor for further cracking.

It is anticipated that a commercial plant could be on-stream in 1980.

(Refer to "Garrett's Coal Pyrolysis" in the Liquefaction Section of this report for additional information.)

GEGAS

Process Developer: General Electric Research and Development Centre, Schenectady, N.Y.

Sponsor: General Electric Company, New York, N.Y.

Description: The process employs a moving fixed-bed gasifier however trouble is experienced with caking coals. To overcome this problem, inert bulk diluting agents such as silicon carbide or coal ash are utilised thus increasing mass-flow through the gasifier. An extrusion process is used for coal-feeding and off-gases are cleaned of hydrogen sulphide using liquid selective membranes. A methanation step is required to up-grade the raw gas to pipeline quality.

Status: Preliminary tests have been completed in a 50 lb. per hour unit. General Electric is seeking partners for erection of a demonstration plant.

HYDRANE

Process Developer: U.S. Bureau of Mines, Pittsburgh Energy Research Centre, Pittsburgh, Pa.

Description: Crushed raw coal is fed to a two-zone hydrogenation reactor operated at 1,000 p.s.i. and $1,650^\circ$F. In the top zone the coal falls freely as a dilute cloud of particles through a hydrogen-rich gas containing some methane from the lower zone. About 20% of the raw-coal carbon is converted to methane, causing the coal particles to lose their volatile matter and agglomerating characteristics. The coal is now essentially a char. This char falls into the lower zone where hydrogen feed-gas maintains the particles in a fluidised state and also reacts with about 34% more of the carbon to make methane. The product gas exits from the centre of the reactor and is cleaned of entrained solids and some unwanted gases. After clean-up, methanation of the small amount (2 to 5%) of residual carbon monoxide gives a pipeline-quality, high BTU gas. Char from the lower zone of the hydrogasifier is reacted with steam and oxygen to generate the needed hydrogen.

Operating Conditions:

Reactor	Temp.$^\circ$F	Press p.s.i.	Reactants	Products
Upper Dilute-Phase	1,650	1,000	Coal-lower-bed gas	Med.BTU Off-Gas
Lower Fluidised-bed	1,650	1,000	Char-hydrogen	826 BTU/CF Off-Gas
Hydrogen Generator	1,800–1,900	1,000	Char-steam-oxygen	Hydrogen-rich Gas

Status: In bench-scale testing, a 10 lb./hour integrated unit has demonstrated the operating feasibility of the process. Results indicate that high-volatile bituminous coals can be fed directly to the gasifier without caking and agglomerating, thereby eliminating expensive pretreatment. Also, 95% of the methane in the final S.N.G. product is made in the hydrogasification reactor from the raw coal directly by treatment with hydrogen. This scheme results in high thermal efficiency (78%).

Scale-up to a 24 ton per day pilot plant is planned. The Bureau of Mines has budgeted U.S. $25.5 million dollars for research and development work on this process for the period fiscal years 1975 to 1979. This will probably be increased.

HYGAS

Process Developer: Institute of Gas Technology, Chicago, Illinois, an affiliate of the Illinois Institute of Technology.

Sponsors: American Gas Association and Office of Coal Research.

Contractors: Blaw-Knox (Dravo Corp.), Belmas Corp., and Procon, Inc., a subsidiary of Universal Oil Products.

Description: Coal, pretreated if caking in an air-fluidised-bed, is slurried with light oil and fed into a fluidised drying bed at the top of the gasifier where oil evaporates at 600°F. The fluidised bed moves down to the first stage of the reactor to a temperature zone of 1300 to 1500°F, while char further moves to the second stage to temperatures of 1700 to 1800°F. Hydrogen-rich gas and steam react with char to produce a methane-rich raw gas and residual char, which then passes to a separate gasifier where it reacts with additional steam to produce the hydrogen-rich gas for injection into the two-stage hydrogasifier. Raw product gas, cooled by upward passage through the first stage of the gasifier and drying bed, is collected for removal of carbonisation products and sulphur impurities. Light oil recovered at this point is used for slurry preparation of coal feed. The gas is then fed to a (nickel) catalytic methanation process for additional hydrogenation and up-grading to high-BTU pipeline gas

Operating Conditions:

Fluidised Bed	Temp °F	Press.p.s.i.	Reactants	Products
Coal Pretreater*	750	Atmospheric	Coal-air	Fuel Gas,Treated Coal
Slurry Drier	600	1000-1500	Coal-light oil	Slurry oil vaporised
Hydrogasifier-Stage 1	1300-1500	1000-1500	Coal-Stage 2 gas	20% of coal to CH_4
Hydrogasifier-Stage 2	1700-1800	1000-1500	Char-H_2 rich gas	25% of coal to CH_4
Electrothermal Gasifier†	1800-1900	1000-15000	Char-steam	Hydrogen
Steam-Oxygen Gasifier†	1800-1900	1000-1500	Char-steam-O_2	Hydrogen
Steam-Iron Gasifier†	2000	1200	Char-steam-air	Hydrogen

* Lignite and sub-bituminous coals do not require pretreatment.
† Alternative processes for production of hydrogen-rich gas for gasifier.

Status A $10.5 million 75 ton per day pilot plant was constructed in Chicago during 1969-1971 to deliver 1.5 million cubic feet per day of synthetic natural gas. In early 1973, large-scale conversion of coal to S.N.G. was demonstrated for the first time. The necessary design data for a commercial plant using steam-oxygen gasification for hydrogen generation will be available in June, 1975.

Capital and operating estimates for a commercial plant have been made based on pilot-plant studies. It is predicted that commercial plants could be in operation by 1980, producing 250 million S.C.F./day of pipeline gas from 5 millions tons per year of coal from gasifiers approximately 170 ft. tall and 23.5 ft. inside diameter each producing 125 million S.C.F. per day. The estimated cost of produced gas is U.S. $1-15 per 1,000 cubic feet.

Pennsylvania Gas and Water Co. has proposed to the O.C.R. a plan for financing and operating a 5,000 ton per day demonstration plant for Pennsylvania based on the HYGAS process (or a similar process) to produce about 80 million SCF per day of pipeline gas.

48

Hydrogen-production units being considered for the HYGAS process, under the sponsorship of O.C.R., I.G.T./A.G.A. to $18.16 million, are:

1. Steam-oxygen process — hot char from the gasifier reacts with steam and oxygen in a high-pressure fluidised bed at $1800\text{-}1900^{\circ}F$.. This process is currently under evaluation at the Chicago pilot plant.

2. Electrothermal gasification — heat for a char-steam reaction is provided by direct-current heating of a fluidised bed of char. Excess char from this electrothermal reactor is used to produce the electric power required. After batch-testing, a 2-MW electrothermal gasifier was built in 1972 and operated successfully. However, due to increasing power-generation costs, testing of the technique has been deferred. (See "ELECTROFLUIDIC GASIFICATION").

3. Steam-Iron process—residual char from the HYGAS gasifier reacts in a fluidised bed with steam and air to generate a producer (reducing) gas for the steam-iron reactor which delivers a hydrogen-steam mixture back to the HYGAS reactor from the oxidizer. An $18 million steam-iron pilot plant is under construction at the HYGAS pilot plant in Chicago and should be operating by late 1975. Blaw-Knox designed the plant and in October 1973 a contract was awarded to Belmas Corporation for fabrication of the reactors.

Note: l. The "Steam-Iron Process" was initiated at the Institute of Gas Technology in 1961 by Fuel Gas Associates (Consolidated Natural Gas Service Company, Inc., Texas Eastern Transmission Corp. and Consolidation Coal Company) originally interested in hydrogen for hydrogasification of coal, ammonia and petrochemicals manufacture. In 1973, Fuel Gas released all rights to the steam-iron process for producing S.N.G. from coal to I.G.T.

2. Prior to 1964, I.G.T. operated a pilot-scale steam-oxygen slagging synthesis gas unit for operation at atmospheric pressure utilising a wide range of coals.

KOPPERS–TOTZEK

Process Developer:　　Heinrich Koppers GmbH, Essen, W. Germany.

Sponsor:　　Federal Government of the Republic of West Germany.

Description:　　The gasifier is a refractory-lined, horizontal, cylindrical vessel with conical ends. Oxygen, steam and coal react at about atmospheric pressure and $3,300^{\circ}F$. Fixed carbon and volatile matter are gasified to produce offgas containing carbon monoxide and hydrogen. Coal ash is converted into molten slag a proportion of which drops into a water-quench tank, the reaminder carried by the gas. Low-pressure steam is circulated around burners and refractory to cool them as well as producing process steam. Gas leaving the gasifier is quenched with water to solidify entrained molten ash. After passing through a waste-heat boiler, the gas is scrubbed to remove entrained solids. Scrubbed gas is compressed to 450 p.s.i.g., hydrogen sulphide and a controlled quantity of carbon dioxide is removed by purification. The purified gas is then shifted and methanated, the methanated gas dehydrated and purified to remove carbon dioxide. Dry, pulverised coal of any type may be used.

Operating Conditions:

Reactor type	Temp. $^{\circ}F$	Pressure	Reactants	Product (raw gas)
Entrained fuel	3,300	Atmospheric	Coal-steam-O_2	300 BTU/SCF

Status:	This process was developed commercially by Friedrich Totzek of Essen and the Koppers Co., Inc. (Pittsburgh) in 1949 following U.S. Bureau of Mines demonstration in 1948. There are 16 Koppers–Totzek plants operating around the world none of which are in the U.S.A.. These plants produce a CO/H_2 synthesis gas which is used in the synthesis of ammonia. At the present time, five new K-T plants with a total of 16 gasifiers are under construction. Of all coal-based synthetic ammonia plants erected since 1945, more than 80% are based on K-T gasifiers. A pressurised K-T gasifier is currently being tested in a development project conducted by a joint venture of Heinrich Koppers GmbH and Shell International Petroleum Maatschappij, The Hague, Netherlands. Gasification under increased pressure (450 psi) would not alter the composition of the raw gas from the K-T gasifier but would improve the economy through reduction of power for compression of the raw gas.

An interesting new development of Heinrich Koppers and Imperial Chemical Industries (I.C.I.), Billingham, U.K. is a combined carbon monoxide-shift conversion and methanation unit. This unit's reactor uses a nickel catalyst which simultaneously promotes both the shift and methanation reactions.

Testing of this shift conversion — methanation reactor by I.C.I. on a pilot scale has indicated that commercially acceptable life-times for the catalysts can be expected.

Estimates for a K-T plant producing 250 million S.C.F. per day of S.N.G. (1,000 BTU/SCF) indicate a coal-feed rate of 25,000 short tons per day to a battery of 24 K-T gasifiers (four-head type) in a plant costing U.S. $400 million.

It is intended to construct a K-T demonstration plant in Germany to produce 2.6 million SCF/day of S.N.G. from a coal feed of 145 tons/day. The Koppers—I.C.I. combined shift conversion—methanation reactor will be installed in the demonstration plant to upgrade the K-T gasifier raw gas to pipeline gas.

The Federal Government of Germany is prepared to give considerable financial aid for the construction and operation of this demonstration plant.

In the U.S., Northern Illinois Gas Co. and the State of Illinois are reportedly studying the K-T process, among others, for application at their proposed 80-90 million SCF per day S.N.G. demonstration plant to be completed in Illinois in the early 1980's.

LIQUID-PHASE METHANATION (L.P.M.)

Process Developer:	Chem Systems,Inc., Hackensack, N.J.
Sponsors:	American Gas Association and Office of Coal Research.
Engineering Sub-Contractor:	Olsen Engineering, Houston, Tex. (a subsidiary of Davy Powergas Inc.)
Description:	The process is suited to the conversion of gas containing high concentrations of carbon monoxide and hydrogen (15 –20% CO, 45 – 60% H_2) into methane (CH_4) by affecting the heterogeneously-catalysed reaction of the feed gases in the presence of an inert liquid phase which absorbs the large exothermic heat of reaction as both sensible heat and as latent heat by vaporisation. The reaction proceeds to near completion in a single pass and economics will dictate whether a single-stage reactor only will be used or if an additional polishing reactor will be required in the final design.

In the process, the inert liquid (e.g. mineral oil, C_{15} to C_{21}) is pumped upward through the reactor at a velocity sufficient to both fluidise the catalyst and remove reaction heat. The synthesis gas is passed concurrently upward through the reactor where it is converted to a high-concentration methane stream.

Operating Conditions:

Reactor	Temp $^{\circ}$F	Press p.s.i.	Catalyst	Catalyst life	Reactants	Product
Catalytic	640	500	Ni-230S	2 years	$CO-H_2$	High-BTU Gas

Status:

The project's program is divided into three phases:

Phase I - Bench-scale unit — all testing in a 0.81" x 4' reactor has been successfully completed. This included a 300 hour run.

Phase II - Process Development Unit (P.D.U.) — the P.D.U. was constructed by Artisan Industries, Waltham, Massachusetts. The unit was installed at the research laboratories of Chem Systems. The P.D.U. reactor is 4 inches in diameter by 7 feet high and is designed for 1,500 s.c.f. per hour feed flow. Testing is continuing and a polishing reactor is being added to the L.P.M. process.

Phase III - Pilot Plant — In January, 1974 Chem Systems awarded a sub-contract to a Davy Powergas subsidiary, Olsen Engineering of Houston, Tex. for construction of a skid-mounted pilot plant with a reactor 15' long and 2' diameter to be shipped to a designated site that can supply synthesis gas. The HYGAS pilot plant at Chicago and Consol's CO_2−ACCEPTOR pilot plant at Rapid City, S.D. are leading candidates. The L.P.M. pilot plant should be ready for start-up at one of these synthesis-gas pilot plants in May, 1975. The L.P.M. unit will be capable of processing 2 million S.C.F.D. of synthesis gas to pipeline-quality gas.

Future plans for the L.P.M. process include studying the use of the system to effect both shift and methanation reactions simultaneously.

Note: Conoco Methanation Co. is testing a fixed-bed catalytic methanator at Westfield, Scotland and Catalysts and Chemicals, Inc., is also developing fixed-bed methanation in a pilot plant at Louisville, Kentucky.

LURGI PRESSURE GASIFICATION

Process Developer:

Lurgi Gesellschaft fur Warme — und Chemotechnik mbH, Frankfurt, West Germany.

Licensor:

Lurgi Mineralotechnik GmbH.
Note: The Lurgi group of companies are subsidiaries of Metallgesellschaft A. G. of Frankfurt (Main), West Germany.

Description:

Crushed and dried coal is fed to a moving-bed gasifier where gasification of coal takes place at 350 to 450 p.s.i. Initial devolatisation occurs accompanied by gasification in the temperature range of 1,140 to 1,400°F. Residence time is about one hour. Steam is the source of hydrogen. Combustion of a portion of the char with oxygen supplies the necessary heat. A revolving grate at the base of the reactor supports the fuel bed, removes the ash, and introduces the steam and oxygen mixture. Crude gas leaving the gasifier at temperatures between 700 and 1,100°F (depending on type of coal) contains tar, oil, naphtha, phenols, ammonia, plus coal and ash particulates. Quenching with oil removes tar and oil. Part of the gas passes through a shift converter. Gas from the shift converter is washed to remove naphtha and unsaturated hydrocarbons. Then CO_2, H_2S and COS are removed. The gas is methanated and pipeline gas is produced by final CO_2 removal and dehydration.

For the processing of caking coal, the gasifier is equipped with an additional device which breaks up the coal during heating.

The Lurgi Process can also produce low BTU gas (refer to appropriate section in this report.)

Operating Conditions:

Reactor	Temp °F	Press p.s.i.	Reactants	Product (off gas)
Fixed bed	1,140-1,400	350-450	Coal-steam-O_2	450 BTU/SCF

Status:

The first commercial plant was constructed in 1936 and to date, fourteen plants (comprising 58 units) have been built with further plants, either with or without a methanation step, under construction or in the planning stages. For instance El Paso Natural Gas Co. is constructing a single, full-scale experimental unit to be operational in early 1975. Methanation of Lurgi synthesis gas, using Lurgi's "Hot Gas Recycle (H.G.R.) Process" catalytic methanators, has been performed on a commercial scale at plants at Westfield, Scotland, Sasolburg, South Africa and at a Lurgi test plant in Austria.* Synthesis gas from the gasifier is suitable for use as town gas, synthesis gas for ammonia, methanol, Oxo, and Fischer-Tropsch synthesis for high-purity hydrogen. The SASOL plant in South Africa utilising 13 gasifier units, has been operating commercially since 1954 and plans are in hand for expansion. The U.S. Bureau of Mines has a Lurgi-gasification pilot plant at Grand Forks, N.D. and the Office of Coal Research — American Gas Association sponsors "Lurgi Studies" as part of its R & D program.

* City of Schwechat, near Vienna.

Note: Lurgi GmbH has tested the utilisation of nuclear-reactor waste heat for coal gasification in its pilot plant at Julich, West Germany.

MOLTEN—SALT

Process Developer:

M.W. Kellogg Co., Piscataway, N.J., a division of Pullman Inc.

Description:

Crushed (12 mesh), and dried coal is picked up from lock hoppers by a preheated steam-oxygen stream and fed into the molten-salt gasifier. Recycled sodium carbonate is fed to the gasifier along with the coal. Coal-steam reaction is catalysed by the molten salt contained in the reactor. A gas free of tars is produced at a sufficiently low temperature so that appreciable methane production can also take place.

Use of the molten-salt reaction medium also makes pretreatment of caking coals unnecessary. A fused, cast, high-purity alumina material has been developed as the reactor lining to resist corrosive attack of the molten salt.

A bleed stream of molten carbonate containing the coal ash in solution is withdrawn from the bottom of the gasifier. It is contacted with water to dissolve sodium carbonate. Ash is separated by filtration.

Sodium carbonate solution is carbonated to precipitate bicarbonate. The bicarbonate is filtered out and calcined to restore carbonate which is then recycled to the gasifier.

Raw gas leaving the gasifier at 1,700°F is passed through the heat recovery section. Any entrained salt is recovered. The raw gas is then shifted, purified, methanated and dehydrated to produce pipeline-quality gas.

Operating Conditions:

Reactor	Temp $^\circ$F	Press p.s.i.	Reactants	Product (raw)
Catalytic-molten Na_2CO_3	1,830	420	Coal-steam-O_2	330 BTU/SCF

Status:

In early testing of the process, which was under Office of Coal Research sponsorship from 1964 to 1967, two vessels were used. In the first, steam and coal were reacted in a molten-salt bath to yield synthesis gas whilst in the second vessel, the residual carbon in the circulating molten carbonate was burned with air to reheat the sodium carbonate for re-circulation. Due to problems arising from the corrosive nature of the salt, the small scale testing of this process, under O.C.R. funding, was discontinued.

W.M. Kellogg Co. continued their research and have developed a non-corrosive alumina reactor lining to overcome the corrosion problem and have incorporated the use of a single reactor vessel which not only eliminates circulation of the molten salt, but also simplifies gas handling and reduces further processing requirements. Because of these new developments, the Office of Coal Research again decided to sponsor the work, with special emphasis on the suitability of the process as a method for generating a fuel gas suitable for magnetohydro-dynamic generating systems (M.H.D. systems). This O.C.R. grant was for $145,175 and expired in July 1974.

The process is now being tested in a process development unit ten times the size of earlier vessels in conjunction with an undisclosed company.

MULTIPLE CATALYST
(also known as the "Wyoming Concept" or "The Direct Conversion of Coal to a High-BTU Gas Process".)

Process Developer:

College of Engineering, Natural Resources Research Institute, University of Wyoming, Laramie, Wyo.

Sponsor:

Office of Coal Research.

Description:

This is a method for the direct production of methane from coal and steam with the methanating-catalyst bed placed in the middle of the reactor and heated to the desired temperature by a Lindberg furnace. The temperature is monitored by thermocouples located in the thermo-well inside the reactor. The methane-rich product gas passes through a motor valve which allows the pressure to be controlled.

Operating Conditions:

Reactor Type	Temp $^\circ$F	Press p.s.i.	Reactants	Product
Heated Ni catalyst	1,200-1,300	1,000	Coal-steam	High BTU Gas 850 BTU/scf after purification

Status:

The concept was initially developed at the University of Wyoming and has been supported by the Office of Coal Research under a 5 year $613,000 contract completed on 26/2/1974. Development effort on the direct methanation process consisted primarily of batch-type tests at the bench scale level. The Alliance Research Centre of Babcock & Wilcox in Alliance, Ohio has undertaken the demonstration of long methanation-catalyst life in the direct production of methane from coal with steam in a continuous reactor of one inch inside diameter and a height of four feet. Fifty-five different catalysts were tested in runs from one to 30 hours in duration.

Before larger-scale continuous operation is successful, a better catalyst system must be developed. A combination of an alkali carbonate, such as potassium carbonate, and a nickel catalyst promises to be the most effective.

Note: The U.S. Bureau of Mines has conducted bench-scale experiments on the catalytic reaction of coal and steam at 1900°F in a fixed-bed reactor to produce methane-rich offgas of about 550 BTU/SCF utilising tungsten sulphide as a catalyst.

"SOLUTION–GASIFICATION"

Process Developer: Stone and Webster Engineering Corporation. (now joint venturing with Gulf Oil Corp.)

Description: Coal is slurried in a solvent; then a two-step treatment with hydrogen solubilises the coal and produces pipeline-quality gas without an explicit methanation step. No oxygen is required in the process. The process treats coal as a basic hydrocarbon in which the hydrogen content is increased from 5% in the raw material to 25% in the methane product. Hydrogen for the process is made by reforming part of the product methane with steam. A range of coal types can be used.

Status: The process has been tested in bench-scale runs. In early 1974, the Office of Coal Research awarded a $300,000 contract to the Oklahoma State Department of Industrial Development to evaluate the use of nuclear reactor waste-heat in coal gasification. Assisting the contractor will be Gulf General Atomics Co. (a Gulf Oil subsidiary) and Stone and Webster Engineering Corp. who together plan a demonstration plant, under a new joint venture, wherein a nuclear reactor will provide heat for hydrogen generation for the "Solution–Gasification" process. The O.C.R. contract requires a conceptual design, including estimates of yield, capital requirements and operating costs, for a complete commercial plant for processing about 36,000 tons per day of coal to produce some 600 million cu. feet per day of high-BTU pipeline gas. A 2-year joint R & D program funded by Stone and Webster and Gulf General Atomic costing U.S. $740.000 has been undertaken

Note: 1. Gulf General Atomic carried out a feasibility study in 1969 on use of waste heat from nuclear power stations for coal gasification.

2. Bergbau-Forschung is currently operating an experimental gasifier to process 220 lb./hour of coal to test their gasifier concept which, if successful, will be integrated with a gas-cooled nuclear reactor to supply the heat required for the endothermic steam gasification of coal.

SUN GASIFICATION

Process Developer: Sun Research and Development Co., a subsidiary of Sun Oil Co.

Description: Coal particles are oxidized by molten sodium sulphate in an exothermic reaction. The coal is converted to carbon monoxide, hydrogen and other gaseous products by oxidation of coal with or without steam, by molten sodium sulphate at 1740–2000°F. The sodium sulphide, produced by the reduction of sodium sulphate, is oxidised with an oxygen source to sulphate for recycling to the gasifier. The moisture in the coal increases the hydrogen to carbon monoxide ratio in the product gases. The exothermic reaction results in almost complete gasification of the coal, minimises gas flow, reduces capital investment, and results in less corrosion of the oxide refractories used as reactor linings. The raw gas is suitable for upgrading to pipeline quality.

Status: Sun Research and Development Co. is the holder of U.S. Patent 3,770,399 of 6th November, 1973 pertaining to this process. Details of development work on the process are not available.

SYNTHANE

Process Developer: U.S. Bureau of Mines, Pittsburgh Energy Research Centre, Pittsburgh, Pa.

Contractors: Design — Lummus Company, a Combustion-Engineering, Inc. subsidiary. Construction — Rust Engineering Company.

Description: Coal is first contacted with a steam-oxygen mixture at 800°F in a fluidised-bed pretreater to destroy the caking properties of the coal feed (if any). Decaked coal from the pretreater enters the gasifier at the top and a mixture of steam and oxygen is introduced at the bottom to fluidise the bed. The gasifier operates at pressures up to 1,000 p.s.i. and at a fluidised-bed temperature of 1800°F. Product gas (synthesis gas) leaves overhead and unconverted coal, or char, is withdrawn at the bottom. The char can be burned to generate all the steam required in the process. After removal of tars and solids, the gas passes through a shift converter and acid-gas removal. Finally, the product gas goes to the methanator, increasing the heating value to that of natural gas.

Operating Conditions:

Gasifier type	Temp °F	Press p.s.i.	Reactants	Product (raw)
Fluidised Bed	1800	1,000	Coal-steam-O_2	405 BTU/SCF

Status: The Synthane process has been under development by the U.S. Bureau of Mines since 1961, beginning with laboratory and small-scale (4" dia.) pilot-plant research. Results of this early testing encouraged the Bureau to commission the Lummus Co. to design a 75 ton per day (coal feed) pilot plant. Construction, by Rust Engineering Co. is underway at Bruceton, Pa. with completion expected in mid fiscal-year 1976. The pilot plant is designed to produce 2.4 million S.C.F./day of synthesis gas from the gasifier, which will be converted in catalytic methanators to 1 million S.C.F./day of synthetic natural gas. Two types of methanator will be installed in the pilot plant at Bruceton, a "Tube-Wall Reactor" which employs a technique for spraying Raney nickel onto vertical plates installed inside a cooling tube, and the "Hot-Gas-Recycle" technique which also uses Raney nickel catalysts.

Following successful pilot-plant operation, the Bureau intends to scale-up the pilot plant by a factor of 200 to a commercial Synthane plant costing U.S. $200 million which will process 15,000 tons per day of Pittsburgh coal to produce 250 million S.C.F./day of pipeline gas. The plant, requiring 28 million gallons per day of water, would produce as by-products, 200 tons per day of elemental sulphur and 30,000 gallons per day of benzene, toluene and xylene. Synthane pipeline gas should sell for 80c to $1.00 (U.S.) per thousand cubic feet, depending on the price of feed coal. The fiscal year 1975 budget for the Bureau's coal-conversion research and development is U.S. $57 million with the emphasis on the Synthane process.

Note: The Synthane plant, with minor changes, could be used as a Fischer-Tropsch plant to make motor spirit, as a methanol plant, and as a plant to make low-BTU gas (using air in place of oxygen).

For example, in the Synthane process, as outlined above, the shift converter shifts the H_2:CO ratio to 3:1 for methanation to pipeline gas. For a modified co-product Synthane plant to produce methanol as well as methane, the shift would be 2:1. Comparing the outputs of a pipeline-gas Synthane plant and a methanol co-product operation, we have:

	Synthane	Co-Product
Pipeline gas, million cubic ft/day.	250	185
Methanol, tons per day	0	3,625

TEXACO GASIFICATION

Process Developer: Texaco, Inc.

Description: Coal, or any carbonaceous fuel, and oxygen are reacted in the presence of carbon monoxide and hydrogen at temperatures of 1200-2200°F and pressures of 300-4500 p.s.i., depending on type of feed. Steam may be used optionally. Hydrogen and carbon monoxide produced in the reactor are recycled to the reactor to optimise methane yield. The high-BTU offgas is suitable for upgrading to pipeline quality.

Status: This process has been described in U.S. Patent 3,740,204 of 19th June, 1973, assigned to Texaco, Inc. No details are at hand on development of the process.

Texaco has also investigated the production of synthesis gas from coal in an unpacked reactor by direct partial oxidation with oxygen and steam at temperatures of 1800-3000°F and pressures of 100-3000 psig. The coal is introduced into the reactor in a preheated slurry. This process has been described in U.S. Patent 3,544,291 of 1st December, 1970, assigned to Texaco, Inc.

TOTAL GASIFICATION

Process Developer: Total Energy Corp.

Description: The process employs two integrated gasifiers. Coal is fed to both gasifiers. In the first gasifier, carbon monoxide is formed which is sent to a hydrogen generator,where carbon monoxide and steam react to form hydrogen. The hydrogen stream is sent to the second gasifier where hydrogasification of the coal feed occurs under conditions minimising carbon monoxide formation and optimising methane production. The temperature in the hydrogasifier is controlled by an indirect heat exchanger rather than by the introduction of steam. Thus the process requires only one source of raw materials — coal. Steam for reforming is raised by process heat.

The methane-rich offgas is suitable for upgrading to pipeline quality.

Status: Few details are available on research and development on this process. Sufficient work has been completed to describe the process in U.S. Patent 3,692,506 of 19th Sept. 1972, assigned to Total Energy Corp.

TWO–STAGE FLUIDISED GASIFICATION

Process Developer: Midlands Research Station, United Kingdom

Sponsor: British Gas Council.

Description: Coal is subjected to hydrogenation in two stages: a rapid reaction at 800-850°C and a slower reaction at 900-950°C, at a pressure of about 750 psi. The char produced then passes to a fluidized-bed gasifier operating at 1900°F, producing a lean gas which is subsequently upgraded by catalytic methanation to pipeline quality.

Status: Several fluidised-bed models were produced in order to overcome problems in fluidised char gasification. Recycle rates were determined for different operating conditions utilising a fluidised-bed pressure model. From the results, a design for a pilot plant to produce 4 million SCF/day of synthesis gas utilising a fluidised-bed gasifier operating at 1050°C and 450 p.s.i. has been prepared.

     The British Gas Council (formerly The Gas Council) is the assignee of U.S. Patent 3,607,158 of 21st September 1971 which describes the above process.

WELLMAN–GALUSHA

Process Developer: McDowell-Wellman and Wellman-Galusha.

Description: Crushed coal ($\frac{3}{16}$″ x $\frac{5}{16}$″) dried and fed by an oxygen-steam mixture, is introduced through a revolving grate at the bottom. Gasifiers are available with or without an agitator. The agitator producer has a slowly revolving horizontal arm which spirals vertically below the surfacr of the fuel bed. The agitator reduces channeling and maintains a uniform bed. The temperature of the gas leaving the gasifier is in the range of 1,000 to 1,200°F., depending on coal type. Pressure is about atmospheric. Ash is removed continuously through a slowly revolving eccentric grate at the bottom of the reactor.

     Raw gas is passed through a waste-heat recovery section. Ash, carried over by gas, and tar are removed by scrubbing. The gas is then compressed and shifted. Pipeline-quality gas is produced by purification, methanation and dehydration.

Operating Conditions:

Reactor	Temp°F	Pressure	Reactants	Product(raw gas)
Revolving Grate	1200	Atmospheric	Coal-steam-O_2	270 BTU/SCF

Status: This process has been commercial for over 30 years. Two units are operating in the U.S.A. The units can produce low BTU gas for individual plants, using air instead of oxygen for a fuel gas of 120 BTU/SCF, or synthesis gas using oxygen. The plants in use are small gasification units serving single large industrial plants. The gasifier could serve a complex of smaller plants requiring a high-BTU gas as a substitute natural gas in special cases.

     In 1974, the Office of Coal Research awarded a U.S. $95,000 contract to McDowell-Wellman Engineering Co., Cleveland, Ohio, to determine the feasibility of constructing a fixed-bed gasifier for operation on caking coals. Under the contract, the firm will support its findings with a preliminary engineering design of the gasifier. It would operate at 300 p.s.i. on highly-caking coals with air (low-BTU product) or oxygen (med. BTU product suitable for methane synthesis) moderated with either carbon dioxide or steam.

Note: In February 1975 Applied Technology Corp. announced it had reached agreement in London with Wellman Incandescent Ltd. for exclusive rights to North and South America to Wellman's coal gasification technology.

WINKLER

Process Developers: Davy Powergas, Inc., Lakeland, Florida, a subsidiary of Davy International Ltd., London, and its affiliate, Bamag Verfahrens-Technik GmbH. (W. Germany.)

Description: Crushed coal is dried and fed to a fluidised bed gasifier through a variable-speed screw feeder. Coal reacts with oxygen and steam to produce offgas rich in carbon monoxide and hydrogen. Because of the high temperatures, all tars and heavy hydrocarbons are reacted. About 70% of the ash is carried over by the gas and 30% is removed from the bottom of the gasifier by the ash screw. Unreacted carbon carried over by gas is converted by secondary steam and oxygen in the space above the fluidised bed. As a result, maximum temperature occurs above the fluidised bed. To prevent ash particles from melting and forming deposits in the exit duct, gas is cooled by a radiant boiler section before it leaves the gasifier. Raw gas leaving the gasifier is passed through a further waste-heat recovery section. Fly-ash is removed by cyclones, wet scrubbers and an electrostatic precipitator. Gas is then compressed and shifted. Gas from the shift converter is purified, methanated, dehydrated and compressed to pipeline quality. Thermal efficiency is 75%.

Operating Conditions:

Gasifier Type	Temp.$^{\circ}$F	Pressure	Reactants	Product
Fluidised Bed	1,500-1,800	Atmospheric	Coal-steam-O_2	275 BTU/C.F.*

* Raw gas from gasifier up-graded by methanation to 960 BTU/CF.

Status: This process was developed in Europe over fifty years ago. The process was constructed commercially at 16 plants in a number of countries, using a total of 36 generators. These plants are still operating with the largest having an output of 1.1 million cubic feet per day. The plants produce low BTU fuel gas (with air instead of oxygen) and synthesis gas for the production of methanol, ammonia, and oil by Fischer-Tropsch synthesis.

The last installation was in 1960, however, the process is once again under consideration for current installation, along with Lurgi, Koppers-Totzek and Wellman-Galusha processes. Davy Powergas Inc. is currently developing a high-pressure modification of the Winkler process which should increase the thermal efficiency.

PATENTED LIQUEFACTION PROCESSES NOT AT PRESENT IN A SIGNIFICANT STAGE OF DEVELOPMENT

Patent Holder	U.S. Patent Number	Date	Patent Description
Texaco Development Co.	3,075,912	29.1.1963	Hydroconversion
Texaco Development Co.	3,740,204	19.6.1973	Production of Methane from Carbonaceous Fuel
Transcontinental Gas Pipeline Corp.	3,728,093	17.4.1973	S.N.G. Production

Note: For a more complete listing of major U.S. Patents relating to coal conversion refer to APPENDIX II.

SECTION III

Low - and Medium - BTU
Coal Gasification Processes

AGGLOMERATING ASH

Process Patent Holder: Union Carbide Corporation, New York, N.Y.

Developer: Union Carbide Corp., Battelle Memorial Institute, Columbus Laboratories Columbus, Ohio.

Sponsors: Office of Coal Research and American Gas Association.

Description: Two fluidized-bed systems, a combustor and a gasifier are linked by an agglomerating ash circuit. Coal, and steam react to produce a med. BTU gas. The system is similar to the high BTU process without methanation. Another major feature of the process is the application of the self-agglomerating method of coal combustion to produce a flue gas sufficiently free of entrained particulates for use directly through a gas turbine for power generation. Also, by recirculating hot ash pellets from the combustion vessel to a separate gasification vessel, as is done in this process, there is no need for an oxygen plant to produce a gas free of nitrogen.

Status: The process' near-term potential lies in producing med-BTU gas for high-efficiency (42-44%) combined-cycle power plants. Large combined-cycle plants are operational in France and Germany with operational plants expected in the U.S. in the 1980's.

Union Carbide Corp. and the Montana Power Company applied in December 1973, for O.C.R. funding for a med-BTU gasification plant that would supply fuel for Montana Power Co's Frank Bird Station at Billings. The proposed plant would have a coal-feed rate of 2,000 T.P.D.

At present a 25 T.P.D. pilot plant is under construction at West Jefferson, Ohio, and is expected to be in operation mid 1975.

AVCO ARC—COAL PROCESS

Developer: Applied Technology Division, Avco Corporation, Everett, Mass.

Status: Avco Corp. have been developing this process for a number of years under contract to O.C.R. A sub-contract was awarded to Blaw-Knox Chemical Plants Division of the Dravo Corporation for an evaluation of the commercial feasibility of the Avco Arc-Coal process for the production of acetylene. The evaluation was based on the production of 300 million pounds per year of acetylene. Facilities were included for recovery of by-products such as carbon black, HCN, char, low-BTU fuel gas, and several forms of sulphur. Blaw-Knox reported to O.C.R. on this study in May, 1972.

Also under an O.C.R. sub-contract, Pennsylvania State University undertook studies on the selection of coals for the Avco Hydrogen-Plasma Arc Process for the production of acetylene. Various interim reports have been issued by Penn. State University to O.C.R. on this work.

It is believed that investigation of this process is continuing.

BABCOCK & WILCOX—DU PONT

Developer: U.S. Bureau of Mines.

Description:	Coal and steam are fed to a cylindrical gasifier incorporating primary and secondary reaction chambers operating at atmospheric pressure under slagging conditions. Coal, steam and oxygen react to form a 270 BTU/SCF synthesis gas.
Status:	Following pilot-scale testing by the U.S. Bureau of Mines in the late 1940's, Babcock & Wilcox constructed a small commercial-scale, 5ft. diameter gasifier of similar design for E.I. DuPont at Belle, West Virginia. A 15 ft. diameter unit was later built at Belle and has been operated by DuPont since 1951. This latter plant, at a feedstock rate of 17 ton per hour, produces 25 million SCF/day of 275 BTU/Scf gas.

C.E. ENTRAINED FUEL PROCESS

Developer:	Combustion Engineering Inc., Windsor, Connecticut.
Sponsor:	Office of Coal Research, Consolidated Edison Co., Electric Power Research Institute (Interested)
Description:	This is an atmospheric, entrained-fuel process. Pulverised coal is entrained in air and steam and fed to a gasifier to produce a 125 BTU/Scf offgas. The gasifier operates at temperatures between 3300° and 3400°F. under slagging conditions.
Status:	Combustion Engineering has been awarded a 4 year, $20.6 million contract by O.C.R. for construction, beginning early 1975, of a 120 T.P.D. pilot plant at Windsor, Connecticut to be operational by 1977.
	The overall test programme is funded jointly by Combustion Engineering Inc. and O.C.R. with E.P.R.I. interested in participating. On successful completion of the pilot plant test programmes the next step would be construction of a 200 MW (electric) combined-cycle demonstration plant, presently expected to be operational by 1980.

"COMBINED CYCLE" B & W

Developer:	Babcock & Wilcox Co. Ltd., Research Centre, Alliance, Ohio.
Sponsor:	General Electric Co., Babcock & Wilcox Co. Ltd., Electric Power Research Institute.
Description:	Coal/air is fed to a pressurised, water-cooled gasifier and reacts to form a combustible gas which, after cleaning, is fired in a combustor that discharges to a high-temperature gas turbine, which, in turn, exhausts to the steam generating portion of the cycle to power a steam-turbine generator.
	The gasifier operates at 900°F under high pressure and the combustor at 1600°F and 95 p.s.i.
Status:	Babcock & Wilcox Co. Ltd. and General Electric Co. are jointly sponsoring the development program. The Electric Power Research Institute (E.P.R.I.) is funding the design by Babcock & Wilcox of a gasifier for a planned 480 T.P.D. pilot plant to be installed at a power generation utility site yet to be selected. Previous testing in a 60 T.P.D. plant has been under evaluation since June, 1961 at the Alliance Research Centre.

Note: 1. The same basic process is employed in the "Combined Cycle" processes of the Foster-Wheeler Corp. and Westinghouse Electric Co.

2. Babcock & Wilcox Co's Alliance Centre is also studying the gasification of char with SO_2 in place of oxygen to produce concentrated carbon monoxide.

COMBINED—CYCLE F.—W.

Developer: Foster-Wheeler Corp., Livingston, N.J.

Sponsors: O.C.R.; Foster-Wheeler Corp.; Northern States Power Company; Pittsburgh and Midway Coal Mining Company (Gulf Oil Corp.), and an industry team.

Description: The process utilises an entrained-type, two-stage, slagging pressure gasifier. This gasifier is fashioned after the two-stage BI—GAS process gasifier developed by Bituminous Coal Research, Inc. The low-BTU fuel gas produced by the reaction of coal, steam and air is cleaned and fired in a combustor that discharges to a gas turbine. The hot gas expanded through the turbine is sent to the steam-generating portion of the cycle to power a steam-turbine generator. The gasifier operates at temperatures above $2100^{\circ}F$ and at a pressure of 520 p.s.i.

Status: A research and development program was initiated in August, 1972, co-sponsored by Foster-Wheeler, Northern States Power Company of Minneapolis, Pittsburgh and Midway Coal Mining Co., and the Office of Coal Research. Cost of the overall project is estimated at U.S. $80–$90 million, culminating in the operation of a 1200 T.P.D. demonstration plant to be located at Northern States Power Co.'s Lawrence plant near Sioux Falls, S.D. The power generated will be used in the Northern States Power system. Foster-Wheeler is currently developing a detailed conceptual design of the demonstration plant under a U.S. $8,685,500 contract. Members of the industry team supporting the project are:

Central Hudson Gas and Electric Corp.
Consolidated Edison Company of New York.
Long Island Lighting Co.
New York State Electric and Gas Corp.
Niagara Mohawk Power Corp.
Orange and Rockland Utilities, Inc.
Power Authority of the State of New York.
Rochester Gas and Electric Corp.

These are members of an organisation known as Empire State Electric Energy Research Corporation which is co-sponsoring the program's industry contributions with Northern States Power Co.

Note: This process is essentially similar to the "Combined—Cycle" processes of Babcock & Wilcox and Westinghouse Electric.

CONSOL FIXED BED

Developer: Consolidation Coal Company (Consol), a subsidiary of Continental Oil Company.

Description: A fixed-bed gasifier is fed an improved feedstock consisting of a suitable mixture of coarse caking coal and non-caking pellets of substantially the same size made by pelletising fines, previously separated from the caking coal, in a hot pelletising rotary kiln. This coal feed is gasified conventionally in the fixed bed gasifier without agglomeration problems to produce a low-BTU gas with air or a synthesis gas with oxygen.

Status:

Consol has developed this process to permit the gasification of difficult caking coals. The process is described in U.S. Patent 3,692,505 of 19th September 1972.

ELECTRIC ARC

Developer:

Columbia University, New York.

Sponsor:

Consolidated Natural Gas Co.

Description:

Carbon in the coal feed reacts with steam in a fluid-convection cathode (FFC), high intensity electric-arc at 800-10,000°C. The H.H.V. of the gas product depends on reaction and subsequent quench conditions.

Status:

The research and development of this process is sponsored by Consolidated Natural Gas Co. To date batch tests have been carried out at about 30 KW.

FIXED BED (Kellogg)

Developer:

M. W. Kellogg Co., Piscataway, N.J. a subsidiary of Pullman, Inc.

Description:

This process employs a low-pressure (30–50 p.s.i.) fixed-bed, revolving-grate gasifier to produce a **low** BTU fuel gas (150 B.T.U./scf) with air, and a **medium** BTU fuel gas (300 B.T.U./scf) with oxygen. By-product tar and oil are used as plant fuel or chemical feedstock.

Status:

The Company is presently studying the formation of a Consortium to erect a demonstration plant at an, as yet, unselected utility site.

G.R.D. GASIFICATION

Developer:

Garrett Research and Development Co. La Verne, Calif. a subsidiary of Occidental Petroleum Co.

Description:

This process is similar to the G.R.D. pyrolosis-liquefaction process. A low-pressure step (30-50 p.s.i.) rapidly pyrolyzes coal in the presence of steam and recycled gas. The pyrolyzer also receives partially burned char which supplies heat. The offgas has a H.H.V. of about 600 B.T.U./scf.

Status:

Tests have been carried out in a 50 lb/hr. reactor. The Company is presently seeking support for the construction of a 250 T.P.D. pilot plant.

For further details of this process refer to "Garrett's Coal Gasification" in Section II of this report.

GEGAS

Developer: General Electric Research and Development, Schenectady N.Y.

Sponsor: General Electric Company, New York, N.Y.

Description: This process utilises a moving fixed-bed air-blown gasifier operating at 120 p.s.i. and 1,000^{O}F, and is similar in operation to the high-BTU GEGAS process.

Status: Tested at a 50 lb/hr. rate. General Electric is currently seeking partners for the construction of a pilot plant at a utility site to be selected.

The Office of Coal Research has awarded a $94,853 contract to General Electric to which $133,000 of additional funds will be added by General Electric for construction and operation of a pilot-plant heat exchanger to determine the effects of high-temperature coal gasification products discharging through an MHD generator.

H.R.I. FLUIDISED–BED

Developer: Hydrocarbon Research Inc. a subsidiary of Dynalectron Corp., Princeton, N.J.

Description: Steam and oxygen fluidize a bed with a fuel depth of 25 feet operating at pressures up to 400 p.s.i.g. and temperatures from 1450-1650°F to produce a synthesis gas of 320 BTU/Scf.

Status: Hydrocarbon Research Inc. operated a 26 inch I.D. fluidised-bed gasifier producing ¾ million Scf/day of synthesis gas in the early 1960's. It is believed that H.R.I. have ceased developmental work on this process and are presently evaluating their H.R.I. Gasification Process conceived by A.M. Squires.

H.R.I. GASIFICATION (SQUIRES)

Developer: Concept developed by A.M. Squires, City College, New York.

Sponsor: Hydrocarbon Research Inc. a subsidiary of Dynalectron Corp.

Description: The gasifier incorporates a conical, fluidised-bed where high superficial velocity of the feed stream permits the bed to operate above the ash softening point of the coal (eg. 2,200–2,300^{O}F).

Status: The concept of this process was developed by A.M. Squires and examined by Hydrocarbon Research Inc. H.R.I. has proposed the construction of a 10 T.P.D. pilot plant and it is believed potential sponsors are sought.

    Note: Prior to 1965, H.R.I. experimented on a pilot scale with fluidised-bed gasification.

I.C.I. MOVING BURDEN

Developer: Imperial Chemical Industries Ltd., England.

Description:	Two separate vessels, a gasifier and a combustor are used with a steam fluidised-bed in the gasifier producing a water gas. Char is withdrawn from the gasifier and circulated to the combustor where it is partially burnt with air and recycled to the gasifier to provide the heat for the water gas reaction. This process produces a synthesis gas of approximately 300 BTU/Scf without the necessity of an oxygen plant.
Status:	A large pilot plant was constructed in England by I.C.I. to evaluate the process. However, disappointing results were obtained, primarily associated with the degradation of the char which caused significant char losses from the fluidised-beds as entrained particles in the flue and product gases.
Note:	The process is essentially similar to the Agglomerating-Ash process of Union Carbide/Battelle/Chemico without the emphasis on the agglomerating characteristics of the ash. (see page 59).

I.F.E. TWO—STAGE

Developer:	International Furnace Equipment Company Limited.
Description:	Coal is gasified as it travels down through a fixed-bed reactor which is injected with an air-steam mixture. The ash is removed from the bottom of the reactor via a rotating grate. The offgas has a H.H.V. of 175 BTU/scf.
Status:	This producer is similar to the Marischka gasifier and to a modified (continuous air blowing) I.G.I. Two-Stage cyclic producer. It has been in commercial use for many years.

I.G.I. TWO—STAGE

Developer:	Il Gaz Integrale, Milan, Italy, a subsidiary of Babcock and Wilcox Ltd.
Sponsor:	Offered by Woodall—Duckham Ltd., a subsidiary of Babcock & Wilcox Ltd., Crawley, England.
Description:	The two-stage gasifier consists of a lower, cyclically-operated water-gas generator upon which is superimposed a continuous vertical retort in which the coal is carbonised. Coke or char from the carboniser gravitates to the water-gas generator and is gasified. The usual mode of operation is on a cycle of about four minutes as follows:

The blow gases (air blown) pass through the water-gas generator and up through flues built into and also surrounding the superimposed carbonising zone, and away to be burnt in the steam superheater and combustion chamber before passing through the waste-heat boiler to atmosphere. During this phase the rich gas alone from the carbonising zone continues to flow to the gas treatment plant; the waste gas is prevented from following this path by the operation of a restricting valve on the gas main.

During the run which follows the blow, preheated steam (600—800°C) is introduced into the bottom of the generator. The resultant water gas and steam then pass up through the carbonising charge and through a gas main carrying with them the rich gas produced in the carbonising zone. Most of the heat required for carbonisation is supplied by the water gas and the remainder by the waste gases as described above. The offgas has a H.H.V. of 335 BTU/scf and though the process is air-blown, little nitrogen appears in the product gas.

Status:

The process was originally developed by Italy's Il Gaz Integrale in the 1940's. Other plants of this type which are similar in design are the "Tulley" and "Power-Gas" complete gasification plants. Two stage processes in which the rich gas from the first stage was taken off separately and recycled in order to decompose the hydrocarbons were developed for synthesis gas manufacture. Examples of these are:

Pintsch Hillebrand Process
Koppers Recycling Process
Viag Synthesis-Gas Process
Bubiag Didier Process.

"IN SITU" GASIFICATION
(also known as "UNDERGROUND GASIFICATION".)

Process Developers:

U.S. Bureau of Mines, Energy Research Centres, Morgantown, W. Va. and Laramie, Wyo.

Description:

The method is to drill a series of holes penetrating the coal bed. Hydraulic or explosive shock waves then are induced in the holes to fracture the coal horizontally, thus creating passages for the gasification process and for the resulting gases to escape. To ignite the coal, propane burners are inserted in the holes, lit, and air pumped down to react with steam (from underground moisture) to form low BTU gas. The gas is created all along the fracture planes and vented to the surface by a pipe system. Thermal efficiency is about 75%.

Status:

The U.S. Bureau of Mines first experimented with the process in the 40's and 50's without much success at Gorgas, Ala. Since then, improvements have been made in such areas as explosive fracturing, fracture orientation and directional drilling. Each has potential merit for improving the technology and economics of the process and the Bureau has resumed laboratory and field tests. The new field tests are underway near Hanna, Wyo., in a coal bed 25 to 30 feet thick under 400 feet of overburden. The Rocky Mountain Energy Co., a subsidiary of Union Pacific Corp., has a pact with the Bureau for testing of this seam.

The Hanna project got underway in March 1973 and results were first published in December 1973. Steady-state operating conditions lasted from September 1973 to February 1974 with a gas production rate of 1.6 million cubic feet of 130 BTU per cubic ft. gas per day. Researchers say they had complete control of flame-front movement so long as they used counter-current air injection. The Bureau hopes to design a small demonstration plant.

The Bureau has signed an agreement with Consolidation Coal Co. and Continental Oil Co. to conduct "In Situ" gasification in the Grants District of Wetzel County, West Virginia. This covers a 5-year program at an expected cost of U.S. $10 million. The University of Texas, Austin is studying the process for gasification of Texas lignite under a $100,000 contract with Mobil, Conoco, Texas Utilities Services Co. and NSF–RANN. Laboratory and process development units will be built followed by field testing.

"In Situ" gasification has also been tried in the U.S.S.R., beginning in 1934 and reaching commercial operation in several plants near Moscow. It is believed that currently interest in underground gasification in the Soviet Union is waning. In the U.K., underground gasification was first suggested by Sir William Siemens in 1868. The C.E.G.B. is now conducting pilot-scale trials in a 5Mw plant at the Newman Spinney Station of the N.C.B. Large-scale trials have also been conducted in Belgium, Czechoslovakia, Italy, Poland and Morocco.

Note: 1. U.S. Bureau of Mines' "Project Thunderbird": Feasibility study undertaken 1967–1969. "In situ" gasification of 50 ft. seams 1,000 to 2,200 ft. deep (uneconomic to mine) investigated including the use of nuclear explosions for seam fracturing. Lean gas produced underground could be processed to liquids and gases on surface. Promising economics.

2. Arthur D. Little, Inc. also evaluated the underground gasification process in 1972 and reported that it could be technically feasible.

3. The U.S. Atomic Energy Commission has patented (U.S. Patent 3,794,116 – 26th February, 1974) a process for the "In Situ" gasification of deep (600–3000 feet) coal seams. The process involves fracturing the coal seam, by nuclear explosions if suitable, introducing a mixture of oxygen and fuel gas, heating the upper layer of the fragmented coal to reaction temperature and recovering the product gas for processing to a high–BTU gas on the surface. Operating pressures are carefully monitored to optimize methane formation and to prevent groundwater seepage. Withdrawal of the product gas is balanced with the reactant input to maintain the operating pressure.

4. Cities Service Oil Co. holds U.S. Patent 3,775,073, 27th November, 1973, which describes an "In Situ" gasification process which utilises a radially-extended horizontal fracture pattern, created by hydraulic fracturing, connecting two or more wells in a coal seam. The seam is subjected to an excess of a first-combustion-supporting gas at a pressure greater than the overburden pressure, so as to distribute the combustion-supporting gas throughout the seam. Subsequently, the coal seam and first-combustion supporting gas are ignited while simultaneously preventing any fluid or gas production from the seam so as to form a network of crumbled coal within the seam. Injection of a second combustion-supporting gas into one or more wells and production of gas and tar liquids from one or more of the wells completes the process.

Cities Service Oil Co. also holds U.S. Patent 3,770,398 of 3rd November, 1973 which describes an improved "In Situ" process for producing high-BTU gas by optimising the water-gas shift reaction by injecting into the reaction zone a mixture of carbon dioxide and steam. The temperature of the reaction zone in the coal seam should be above $1800^{o}F$. The production of high-BTU gas would require methanation of the synthesis gas on the surface.

5. West Virginia University has been let two contracts totalling U.S. $29,983 by the Bureau to carry out computer studies on "In Situ" gasification.

6. E.R.D.A. (formerly O.C.R.) has let a U.S. $676,000 contract to Sandia Laboratories, Albuquerque, N.M., to monitor the Hanna experiments. It is expected that this research will support the high BTU "In Situ" gasification project being conducted by the Lawrence Livermore laboratory, Livermore, California.

7. The Alberta and British Columbia governments of Canada are studying the feasibility of developing "In Situ" coal gasification projects within their provinces.

8. Texas Utilities Services Co. and the Soviet Union's Scientific Research Agency have recently contracted to make Soviet technology available for gasification of deep lignite seams in Eastern Texas. The low-BTU gas would be used by Texas Utilities members Dallas Power and Light Co., Texas Electric Service Co., and Texas Power and Light Co. as an alternate energy source. A contract has been signed with V/O Licensintorg of Moscow for the sale of the Soviet technology for less than U.S. $2 million. The Soviet Union is to get "modest royalties" on any gas developed by the process. Texas Utilities will build a pilot plant as the first step in the new project.

9. An E.R.D.A, 5-year contract has been let to the University of California to conduct testing in the Powder River Basin coalfield in Wyoming. A further program is planned for southern Wyoming.

KERPELY PRODUCER

Developer: U.S. Bureau of Mines (operator).

Description: Coal passes through a lock hopper down into a fixed-bed cylindrical unit where it is gasified by a steam-oxygen (or air) blast through a revolving grate which removes the ash continuously. The unit operates at atmospheric pressure to produce a 260 BTU/SCF gas with oxygen or a 130 BTU/SCF gas with air.

Status: A 7 feet internal diameter unit was operated by the U.S. Bureau of Mines research station at Louisiana, Missouri, producing about 2 million SCF/day of 260 BTU/SCF synthesis gas with oxygen blasts.

KOPPERS—TOTZEK

Developer: Heinrich Koppers GmbH; Essen, W. Germany.

Description: Pulverised coal along with oxygen and steam react in a high-temperature, (3,300°F) atmospheric-pressure gasifier to yield a 300 BTU/scf product gas. Any type of coal, irrespective of properties can be gasified almost completely and instananeously.

Status: This process is essentially similar to the Koppers—Totzek high BTU gas process differing in that the product gas is not shifted and methanated. Currently there are 16 plants in operation producing a synthesis gas for the production of ammonia and methanol. The Koppers—Totzek plant is currently being considered for installation as the char-gasification unit for the existing COED pilot plant at Princeton N.J.

For additional details on this technique refer to Section II of this report.

LASER IRRADIATION PYROLYSIS

Description: The Laser irradiation of coal pyrolyses the coal rapidly at high temperatures to produce a gas containing acetylene, hydrogen, carbon monoxide and carbon dioxide. No liquid product or significant methane are produced due to the high temperatures generated by Laser energy. The gas yield varies inversely with coal rank.

Status: Experiments have been conducted by the U.S. Bureau of Mines to investigate Laser pyrolysis of coals of various ranks. The gaseous products of pyrolysis were analysed by mass spectrometry.

Tests were also conducted to examine the effect on the Laser pyrolysis of coal of argon, nitrogen and helium atmospheres. Results indicated optimum generation of acetylene under argon pressure.

The U.S. Bureau of Mines and Georgia Institute of Technology have also studied the reaction of coal in argon and argon-hydrogen plasma jets at temperatures of 3,400-12,000°C to produce acetylene. (See also "Avco Arc-Coal Process" in this section.)

LURGI

Process Patent Holder:	Lurgi Gesellschaft fur Warme-und Chemotechnik mbH. Licensed by: Lurgi Mineralotechnik GmbH.
Developer:	Lurgi (which is a subsidiary of Metallgesellschaft A.G. of Frankfurt (Main) West Germany).
Description:	Coal is fed via a lock-hopper to a gasifier operating at about 350 p.s.i. A revolving grate at the base of the gasifier removes ash and allows air and steam to enter. The product gas has a H.H.V. of about 185 BTU/scf. For additional details on this process refer to high-BTU section.
Status:	This process has been utilised commercially in many plants over a number of years to produce synthesis gas, fuel gas and town gas.

Chemotechnik & Steag carried out studies into combined gas/steam cycle power generation at a 170 MW unit built in a linear power station, comprising a 74 MW gas turbine combined with a 96 MW steam turbine utilising a modified Lurgi gasifier for power in Lunen, Germany. Recently, expansion of the plant to 800 MW was announced.

In the U.S. Commonwealth Edison and E.P.R.I. announced in early 1974 construction of a $19 million low BTU gasification unit at the Powerton Generating Station near Pekin, Illinois. The plant is scheduled for operation in 1977 converting 60 T.P.H. of coal into clean fuel (190million S.C.F./day) for a 70 MW generator. The installation will utilise three Lurgi gasifiers.

The U.S. Bureau of Mines is also conducting studies on the first Lurgi pressure-coal gasifier installed in the U.S. by Blaw-Knox at the Pittsburgh, Pa. centre and has a pilot-scale Lurgi slagging gasifier at Grand Forks, N.D. for operation on lignite.

In the United Kingdom, both the British Gas Council and the B.C.U.R.A. have conducted pilot-scale tests on Lurgi slagging gasifiers, blown with a steam-oxygen mixture, to produce a med-BTU synthesis gas.

MARISCHKA

The Marischka coal gasifier has been in commercial operation for many years, mainly for the gasification of anthracite or coke. In a 6' 6" diameter gasifier production is of the order of 2.5 to 3.5 million SCF per day.

Special features of the gasifier are the annular boiler with upper and lower sections connected by two closely-spaced concentric rows of water tubes within the reactor. The hot gases leaving the gasifier pass into a chamber external to the lower steam jacket and leave near the base of the vessel. The steam raised at 100 p.s.i., amounts to 1.0 to 1.4 pound per pound of fuel gasified.

No information is available on the current status of the system.

MOLTEN—SALT

Developer:	M.W. Kellogg Co. Houston, Texas, a division of Pullman Inc.
Sponsor:	Office of Coal Research.
Description:	This process is essentially similar to the high-BTU "Molten-Salt" process differing in the use of air instead of oxygen thus producing a lower BTU gas of 150 BTU/scf.
Status:	The Company is presently studying the formation of a Consortium to erect a demonstration plant at an, as yet, unselected utility site.
	M.W. Kellogg Co. has also investigated combined desalination and power generation using coal energy under an O.C.R. contract, and fixed-bed coal gasification. (Refer to "Fixed-Bed" process in this section of the report for additional information.)

MULTIPLE FLUIDISED BED

Developer:	Bituminous Coal Research Inc. (an affiliate of The National Coal Association)
Sponsor:	Office of Coal Research.
Description:	Multiple fluidised beds are employed in a gasifier to produce a gas free of liquids. Air is used to produce a low BTU offgas of 160 BTU/scf. This process is similar to the high-BTU BI—GAS process.
Status:	Blaw-Knox Division (Dravo Corp.) was awarded a contract for the engineering, procurement and construction of a 100 lb/hr. process engineering development unit (PEDU) to be constructed at Monroeville, Pa. Construction commenced in 1974, the work being sponsored under an O.C.R. contract of $2.75 million for research and development of the suitability of the process for the production of low-BTU fuel gas.
	(Refer to "BI—GAS" process in the High—BTU section of this report.

OTTO RUMMEL SLAG BATH (Double Shaft)

Developer:	Dr. C. Otto and Co.
Description:	This process is similar to the Rummel Slag Bath (single shaft) process. In this case, however, the exothermic air-blast phase and the endothermic water gas phase are applied to separate sections of a common slag bath produced by means of a vertical partition reaching a short distance into the bath. Thus, synthesis gas relatively nitrogen-free can be generated using air to produce a

69

gas of 270 BTU/SCF. Excess slag is continuously withdrawn via an overflow weir located in a central annulus.

Status:

In London, the Gas Council Research Station conducted tests in a 3 ft. I.D. pilot plant unit from June 1962, however experimental results caused rejection of the pilot plant design; primarily problems with iron separation from the slag were experienced.

Note; The Kellogg molten-salt process is analageous to this process however molten sodium carbonate is utilised to remove the ash from the slag bath.

PANINDCO

Description:

Pulverised coal (-200 mesh) is fed to the centre of a refractory-lined cylinder with a domed top. Oxygen-steam or air-steam mixtures are fed into an annular space surrounding the coal feed. Steam is fed through several nozzles where it is used as a gasifying medium and to moderate reaction temperatures and protect the refractory lining. Ash and product gas are removed from the bottom of the vessel. With oxygen a synthesis gas of 210 BTU/SCF and with air a gas of 125 BTU/SCF are produced.

Status:

The process has been tested on a pilot scale in an experimental plant processing 1,600 lb/hr. of coal feed in Rouen, France which was installed in 1950.

PHILADELPHIA AND READING

Developer:

The Philadelphia and Reading Corporation, New York.

Description:

No details are available on the process.

Status:

The Philadelphia and Reading Corp. became interested in the possibility of utilising pressure gasification of anthracite culm and silt to produce a wide range of products, including hydrogen, fuel gas and ammonia. They retained Blaw-Knox Chemical Plants Division of the Dravo Corporation to provide designs and definitive estimates of costs for producing 100 million SCF per day of 98% pure hydrogen from the anthracite material. The hydrogen could then be used for the synthesis of ammonia, nitric acid, ammonium nitrate, urea, methanol, formaldehyde and pipeline gas.

Dorr-Oliver, Inc. have also investigated fluidised-bed combustion and gasification of anthracite culm-bank material under an O.C.R. — sponsored project called "Operation Red Dog" which was finalised in November, 1969.

PINTSCH HILLEBRAND

Description:

Coal is subjected to primary distillation in an upper chamber and to gasification in a lower chamber. Distillation gas is recycled to regenerative heaters coupled with a producer gas generator for gas and steam heating and these supply the steam and heat to the gasification zone where water gas is produced. The gas has a heating value of 280 BTU/SCF.

70

Status: This process has been in commercial use in Germany for many years and a 15 ft. I.D. reactor produces approximately 4.9 million SCF of synthesis gas per day.

POWER-GAS

Developer: Power-Gas Co.

Description: Coal is fed to the top of the gasifier where it is gasified at atmospheric pressure with air to produce a 160 BTU/SCF offgas. The ash is continuously removed from the base of the vessel.

Status: This process has been in commercial operation for many years.

RAPID, HIGH TEMPERATURE

Developer: Eyring Research Institute, Provo, Utah.

Sponsor: Office of Coal Research.

Status: The Office of Coal Research (O.C.R.) has awarded a $208,000 contract to E.R.I. for research on a rapid, high temperature process to convert coal to a clean fuel gas. The research will continue work previously supported at Brigham Young University under a sub-contract from Bituminous Coal Research Inc.

RILEY-MORGAN

Developer: Riley Co.

Description: Coal is fed to a low-pressure gasifier utilising a fixed-bed system. The H.H.V. of the product gas varies with the use of air (low) or oxygen (med.).

Status: This process can be utilised to produce industrial fuel gas and updated versions are offered with Morgan Construction Co. gasifiers.

ROCKGAS

Developer: Atomics International, a division of Rockwell International Corp.

Sponsor: Office of Coal Research.

Description: Coal is contacted with air in molten sodium carbonate at $1,800^{\circ}F$ and 75 to 150 p.s.i. Sulphur and ash are removed externally from a recirculating bleed stream of the salt.

Status: This process has been tested in an atmospheric 200 lb/hr. unit. The Office of

71

Coal Research have awarded Rockwell a 40 month, $6.9 million contract to demonstrate the feasibility of the process to produce a low BTU gas for electrical-power generation and to design build and operate a 120 TPD pilot plant demonstration plant at the Norwalk Harbour plant of Connecticut Light and Power Company.

RUHRGAS VORTEX

Developer: Ruhrgas A.G., West Germany.

Description: High-ash coal or lignite crushed to a $1/16''$ top size is introduced with air, preheated to 1300^0F, into a vortex chamber where it is gasified without steam and under slagging conditions. The slag passes out the bottom and the reactants pass upward into a tall shaft of larger diameter where the greater part of the gasification occurs at 3,100^{0}F. The 100–120 BTU/SCF offgas is passed through cyclones and bag-filters for dust removal and recycling of entrained char.

Status: This process has been in commercial use for many years to produce low-BTU fuel gas.

RUMMEL SLAG BATH (Single Shaft)

Developer: Union Rheinische Braunkohlen Kraftstoff A.G. Wesseling, West Germany.

Description: Fuel particles in suspension and gasifying medium are injected through twelve tangential nozzles into a slag bath maintained in the base of the producer shaft. Reactants and slag are bought into intimate contact where the coal particles are entrained in the slag allowing for a high conversion of coal to gas and of ash to slag.

The slag continuously overflows and is quenched in water, offgases are cooled in the top of the reaction. The gasifier can be operated with air or oxygen to produce a 110 BTU/SCF fuel gas or a 270 BTU/SCF synthesis gas.

Status: The process has been commercially available for many years. A 6 ft. diameter reactor with a capacity to produce 13 million cu.ft. per day of synthesis gas has been commercially operated in Germany.

STIRRED FIXED–BED
(also known as the "MORGAS" process)

Developer: U.S. Bureau of Mines, Pittsburgh Energy Research Centre, Pittsburgh, Pa.

Sponsor: Office of Coal Research and U.S. Bureau of Mines.

Description: The process employs a fixed-bed gasifier equipped with a stirrer to break up any coke that is formed in the upper section. The stirrer both rotates on its shaft and moves vertically in the reactor. Two coal hoppers feed coal to the top of the gasifier. Steam and air are introduced at the base of the bed through a revolving grate. Ash is removed from the gasifier by the revolving grate to an ash pit. The unit is applicable to the production of S.N.G. by adding shift and methanation steps as in the SYNTHANE process.

Status: Unlike most other fixed-bed gasifiers, it is possible to gasify highly-caking coals using the stirrer to agitate the bed. McDowell-Wellman Engineering has been awarded an $94,919 contract by O.C.R. to design a stirred,fixed-bed gasifier similar to this for operation on caking coals. An 18 T.P.D. pilot plant is in operation at the Bureau of Mines, Morgantown, W.Va., Research Centre. Additional research on high-pressure operation is continuing.

Note: O.C.R. and U.S. Bureau of Mines are also co-sponsoring a current, 3-year R & D program on a combustion system that will provide a hot, clean working fluid from coal gasification in a 3-stage, high temperature combustor which is suitable for magnetohydrodynamic power generation. Design of a 1,000 pound per hour pilot plant was completed in 1973. It is planned to build and operate this pilot plant and investigate the performance of this combustion-MHD system.

TEXACO GASIFICATION

Developer: Texaco Development Company

Description: Pulverised coal in suspension in steam at approximately $950^{\circ}F$, is fed into the top of a downflow cylindrical reactor where it is gasified with oxygen to produce a synthesis gas of about 280 BTU/SCF.

The unit operates under a pressure of 260 p.s.i. and temperatures above the ash fusion point with the product gas and slag being removed from the base of the vessel.

Status. This is a sub-commercial process tested in a down-flow plant and has also been utilised in the I.G.T. Hygas pilot plant in the early days of operation for the production of hydrogen from residual char. It was replaced by the steam-iron process coupled with a Winkler gasifier.

THYSSEN GALOCSY

Description: Coal is fed to a fixed-bed cylindrical gasifier where reactants are injected at three levels. Recycle gas, oxygen and steam in the basal section and oxygen at the two higher levels.

The unit operated at atmospheric pressure and temperatures above the ash fusion point to produce a synthesis gas of about 320 BTU/SCF.

Status: Commercial scale generators were built, typically of 10 ft. internal diameter and 36 ft. in height, producing approximately 3 million SCF/day.

TWO—STAGE PROCESS
(also known as "SUBMERGED COAL COMBUSTION")

Developer: Applied Technology Corporation, Pittsburgh, a subsidiary of International Systems & Controls Corp.

Sponsors:	American Gas Association and the Office of Coal Research (1973); Environmental Protection Agency (1974).
Description:	The process is based on the molten-iron gasification process with air/coal feed to yield a 185 BTU/SCF offgas, and is essentially similar to ATGAS/PATGAS.
Status:	This carbonisation process is common to all three of the molten-iron coal gasification processes i.e. ATGAS, PATGAS and TWO-STAGE, however,the former two processes are oxygen blown. Testing has been carried out in short duration runs using a 2 ft. internal diameter vessel.

The U.S. Environmental Protection Agency is sponsoring a design study for a 50 to 100 MW power generating plant utilising the process' low-BTU offgas.

TWO–STEP COAL PYROLYSIS–GASIFICATION PROCESS

Developer:	West Virginia University, Chemical Engineering Department, Morgantown, W.Va.
Sponsor:	Office of Coal Research.
Description:	Coal, including caking coal, is fed to a sand fluidised-bed for pyrolysis at 1400°F. Char produced is separated from the effluent gas and reacted with air and steam in a gasifier to produce fluidising gases for the pyrolyser, adding a small quantity of raw coal to the char feed to the gasifier to maintain the temperature of the gasifier at 1900°F and to produce enough gases to fluidise the incoming coal in the pyrolyser.

The offgases from the two stages are combined and purified to produce a low-BTU gas for power generation.

Operating Conditions:	Reactor	Temp.$^{\circ}$F	Press.p.s.i.g.	Reactants	Product
	Pyrolyser	1400	10	Coal-heat	Pyrolysis Gas
	Gasifier	1900	10	Char-air-steam	Low-BTU Gas

Status:	West Virginia University's Chemical Engineering Department, under O.C.R. sponsorship, has demonstrated in bench-scale experiments the feasibility of the process. Experiments were conducted in a 15 inch-diameter fluidised-bed reactor filled with 0.025 inch-diameter sand to a collapsed height of 30 inches to test the pyrolysis stage of the process. The second-stage gasification step would employ conventional steam-air gasification.

A conceptual study has been made of the process for application in power generation by coupling the Two-Step process with an advanced-design combined gas and steam turbine power cycle. The conceptual scheme is estimated to generate electricity with a higher efficiency than with a single-step coal gasifier system.

U–GAS

Developer:	The Institute of Gas Technology (I.G.T.) Chicago, Illinois.

Sponsor:	American Gas Association,Office of Coal Research.
Description:	Crushed coal (pretreated if caking) is pressurised in a lock-hopper operating at 350 p.s.i. and 800°F., reacted with air then fed to a fluidised-bed gasifier operating at 350 p.s.i. and 1,900°F. Air and steam are introduced to the base of the gasifier and ash removed through lock-hoppers. Gases from the preheater and gasifier pass through heat-recovery and sulphur removal systems and power recovery turbines reducing the pressure to desired limits. Gas produced is 155 BTU/SCF. Substituting oxygen for air produces a med. BTU fuel gas.
Status:	The research and development of this process is co-sponsored by AGA/OCR. The current programme has tested the process in an air-blown 485 lb/hr. unit showing the suitability of the process for both combined-cycle power generation and as a "grass-roots" source of industrial and power-generation energy. A design study is presently underway for a 10 to 35 TPH pilot plant sufficient to fuel a 100 MW power utility.

Conceptual studies for a 1000 MW commercial combined-cycle power generation plant indicate a coal feed of 7,350 T.P.D. to three gasifiers each 22 feet in diameter and 30 feet high. Thermal efficiency should be greater than 43%.

U.G.I. BLUE WATER GAS

Developer:	U.G.I. Corporation.
Description:	In this process, a producer containing a bed of coke is steam-blasted in the make phase, during which the gases produced are passed off and held in.a refief holder. In the blown-phase, air is blown through the bed to restore the temperature to that before the water-gas reaction. The nitrogen-rich gas from this phase is not stored with the synthesis gas produced by the steam blast. In this way, a 295 BTU/SCF synthesis gas is cyclically produced.
Status:	This process has been in commercial use for many years. The process has been modified for continuous operation with steam and oxygen for production of 270 BTU/SCF synthesis gas by E.I. Du Pont de Nemours Company in a commercial plant.

Other processes similar to this modified Blue Water Gas producer are the Thyssen Galoczy Synthesis Gas process, the Leuna Synthesis Gas process and the Kerpely Synthesis Gas process, all of which produce a 250 BTU/SCF synthesis gas by using steam and oxygen blasts over coke beds.

WELLMAN–GALUSHA

Developer:	McDowell – Wellman Co., Cleveland, Ohio. Wellman – Galusha Co., England.
Description:	Coal, steam and air are fed to a revolving grate gasifier operating at atmospheric pressure and 1,200°F, to produce a 120–160 BTU/SCF fuel gas. For additional details on this process refer to High-BTU section.
Status:	This process has been in commercial use for over 30 years. The plants produce either a fuel gas (air) or a synthesis gas (oxygen).

The yield of gas per ton of coal or coke gasified varies with the moisture and ash contents of the fuels fired. Under good operating conditions with bituminous coals of weakly-caking types and containing about 10% moisture and 6% ash a yield of 140,000 to 150,000 cubic feet of 160 BTU/SCF gas can be considered typical Wellman gasifiers in use vary from 8 feet to 11 feet in internal diameter, the largest size having a capacity of 55 T.P.D. of coal with a gas output of about 8 million S.C.F. per day. Because of the small size of the reactors, the vendors can only offer units suitable for single, large industrial plants or a complex of smaller plants, requiring fuel or synthesis gas.

Wellman Incandescent Ltd. is offering a modified two-stage producer-gas process, similar to the Wellman mechanical gas producer, which produces a hot, detarred fuel gas. 20 two-stage producers are in operation or on order.

Note: In North and South America, Wellman gasifiers are offered by Applied Technology Corp., Pittsburgh, Pa., under an agreement with Wellman Incandescent Ltd. of London.

WESTINGHOUSE FUEL CELL

Developer: Westinghouse Electric Corp., Research and Development Centre, Pittsburgh, Pa.

Description: Coal is fed to a gasifier operating at about 1,800°F to produce gas which energises solid-electrolyte fuel cells immersed in the gasifier bed. Water vapour and carbon dioxide from the fuel cells react with the coal during gasification.

Status: This concept was studied by Westinghouse from 1962 to 1970 under an O.C.R. — sponsored R & D program known as "Project Fuel Cell". Jackson and Moreland Division of United Engineers and Constructors, Inc. provided review and evaluation reports on the project under an O.C.R. contract.

The concept was found to be attractive however no project is now current.

WESTINGHOUSE LOW—BTU PROCESS

Developer: Westinghouse Electric Corporation, Research and Development Centre, Pittsburgh, Pa.

Sponsors: Bechtel, Inc., AMAX Coal Co., Peabody Coal Co., Public Service of Indiana, Westinghouse Corp., Office of Coal Research (O.C.R.)

Description: Crushed (¼" x 0), dried coal is fed into a central draft tube of the devolatiliser—desulphuriser unit (gasifier). Coal and internally—recycled solids are carried upward in the draft tube by hot gases from a combustor flowing at a velocity greater than 15 f.p.s. Recycle solids flow downward in a fluidised bed surrounding the draft tube at rates up to 100 times the coal feed rate. They dilute the coal feed to prevent agglomeration as it devolatises. Heat requirements of the coal-steam gasification reactions are provided by hot gases produced in the combustor. A lime sorbent is added to the devolatiliser—desulphuriser reactor to remove sulphur which is present as hydrogen sulphide in the gas. Spent sorbent is withdrawn from the reactor after stripping out the char. Spent sorbent is regenerated and recycled to the reactor. Char is withdrawn from the top section of the devolatiliser—desulphuriser and fed to the combustor. Char is gasified with air and steam at 2,100°F. Ash agglomerates at the temperature of the combustor and is removed. Raw product gas (135 BTU/SCF) from the devolatiliser-desulphuriser unit passes through a cyclone to remove fines and then through a heat-recovery unit. Fines are recycled to the combustor.

Status: This process is being tested in a 1200 lb./hour pilot plant at Waltz Mill, Pa. Westinghouse in late 1972 began a 9-year research and development program, expected to cost U.S. $80 million, co-sponsored by Bechtel Corp., AMAX Coal, Peabody Coal Co. and the Public Service of Indiana.   Eleven electric power utilities* are also sponsoring the program as associate members. A 60 ton per hour commercial, low—BTU gasification and electric power plant is under construction at the Dresser Station of the Public Service of Indiana at Terre Haute, Indiana. This project, also sponsored by O.C.R., will utilise a combined-cycle coal gasification—power generation system fueled by the Westinghouse gasifier. The plant is expected to be operational during 1978.

In the 1980's it is hoped to operate power stations of 500MW capacity and 50% thermal efficiency without pollution from sulphur or nitrogen oxides, or particulates, at costs well below those of conventional plants.

> * Northern Indiana Public Service Company
> Tennessee Valley Authority
> Consumers Power Company
> Union Electric Company
> Duke Power Company
> New England Electric System
> Columbus and Southern Ohio Electric Company
> Pennsylvania Power and Light Company
> The Montana Power Company
> Tampa Electric Company
> Iowa Power and Light Company

WILPUTTE PRODUCER

Developer: Wilputte Corporation, Murray Hill, N.J.

Description: The gasifier is available in several forms depending on the type of coal used as feedstock. Coal is fed downward to a fixed-bed where it is gasified by partial combustion with moist air passing upwards through the bed.

Ash is withdrawn by a rotating grate at the bottom of the unit.

Status: The gasifier is available with a capacity of 30 ton per day of coal feed to produce 3.5 to 4.0 million SCF/day of 150-170 BTU/SCF producer gas.

WINKLER

Developer: Davy Powergas Inc., a subsidiary of Davy International Ltd., London, and its affiliate Bamag Verfahrens-Technik GmbH.

Description: Crushed coal is fed to a fluidised-bed gasifier where it reacts with air and steam to produce a 120 BTU/SCF offgas. For additional details on this process refer to High BTU section.

Status: This process was developed in Europe over fifty years ago and is used commercially in 16 plants (36 units) in a number of countries. These produce low BTU-fuel gas with air and synthesis gas with oxygen for the production of methanol, ammonia, and oil by Fischer—Tropsch synthesis. Presently, this process along with Koppers-Totzek is being considered for installation in the char-gasification unit at the COED pilot plant at Princeton, N.J.

To overcome fuel limitations associated with the basic Winkler unit, a technique employing a combined fixed bed and fluidised bed operation has been developed. Originally consisting of a single generator (Flesch-Winkler process), this was later modified to operating two generators in parallel (Flesch-Demag process). These combined-reactor systems gasify a wide range of fine coal in a cyclic manner.

REFERENCES CONSULTED

1. "American Gasification Processes Inch Closer to Success", ed. Coal Age, December 1974 (6 pages)

2. "Ammonia Plant for South Africa", Koppers Report, Heinrich Koppers GmbH, Essen,September 1973 (2 pages)

3. "Ammonia Production Based on Coal and Lignite" Koppers Report, Heinrich Koppers GmbH, Essen, October 1971 (2 pages)

4. "Application of COED Process Coal – Derived Liquids in a Petroleum Refinery", Chem Systems Inc. & F.M.C. Corp., 9th Intersociety Energy Conversion Eng. Conf., San Francisco, Aug., 1974 (12 pages)

5. "Aromatics from Coal", K.A. Schowalter and E.F. Petras, U.S. Steel Corporation, American Institute Chemical Engineering Symposium on Aromatics, Houston, Texas, 17-19 March 1975 (18 pages)

6. "Bi-Gas", Hydrocarbon Processing, April 1973 (1 page)

7. "Bi-Gas Process", BCR/OCR/AGA, undated (2 pages)

8. "Charts from Presentation on Coal Gasification", H.L. Malakoff, COGAS Development Co., Princeton (Undated) (23 pages)

9. "Chemical By-products from Coal", Skiest Laboratories Inc., Livingston, New Jersey, 1971 (128 pages)

10. "Clean Automotive Fuel – Laboratory Scale Operation of the Synthane Process", A.J. Forney et alia, Pittsburgh Energy Research Centre, U.S. Dept of the Interior, Feb. 1972 (6 pages).

11. "Clean Boiler Fuels from Coal", J.B. O'Hara et alia, Ralph M. Parsons Co., Los Angeles, Chemical Engineering, June 1974. (13 pages)

12. "Clean-Coke Process for Metallurgical Coke, The", Chem. Eng. Progress, June 1974 (7 pages)

13. "Clean Energy from Coal – New Developments ", Irving Wender, Pittsburgh Energy Research Centre, Bureau of Mines, Pittsburgh June 1974 (15 pages)

14. "Clean Energy from Coal Technology", Office of Coal Research, U.S. Dept. of the Interior, 1974 (43 pages)

15. "Clean Fuels from Coal", W.W. Bodle and K.C. Vyas, Institute of Gas Technology, Chicago, Oil and Gas Journal, Aug. 1974 (16 pages)

16. "Clean Fuels from Coal by the COED Process", F.M.C. Corp., Env: Prot. Agency Sym., St. Louis, May 1974 (20 pages)

17. "Clean Fuels from Coal for Power Generation", John F. Jones, F.M.C. Corp., Am. Chem. Soc., Atlantic City, Sept. 1974 (37 pages)

18. "Clean Power from Dirty Fuels", Arthur M. Squires, Scientific American, undated (10 pages)

19. "Clean Synthetic Fluid Fuels from Coal : Some Prospects and Projections", Elburt F. Osborn, Carnegie Institute, Soc. of Min. Eng., Sept. 1974 (3 pages)

20. "Coal and the Present Energy Situation", Elburt F. Osborn, Science, Feb. 1974 (5 pages)

21. "Coal and Oil Shale Conversion Looks Better", K.C. Vyas and W.W. Bodle, Institute of Gas Technology, Chicago, Oil and Gas Journal, March 1975 (10 pages)

22. "Coal Comes Back", Exxon Corp., New York 1974 (7 pages)

23. "Coal Conversion Processes Loom Big as a Source of Hydrocarbon Fuels", Noel F. Boyd, Dravo Corp., Mining Engineering, Sept. 1974 (8 pages)

24. "Coal Conversion Technology", Harry Perry, Chemical Engineering, July 1974. (15 pages)

25. "Coal Energy – Transport Study", Bechtel Corp., San Francisco, June 1974. (43 pages)

26. "Coal Gasification — A Key Process for Coal Conversion", Paul F.H. Rudolph, Lurgi, A.I.M.E., Dallas, Feb. 1974. (25 pages)

27. "Coal Gasification as a Source of Power", J.W. Loeding & J.G. Patel, Institute of Gas Technology, Chicago, A.I.M.E., Dallas, 1974 (15 pages)

28. "Coal Gasification Technology", Fluor Engineers & Constructors Inc., Feb. 1973. (93 pages)

29. "Coal — It's Changing its Color", April 1974 (3 pages)

30. "Coal Liquefaction at the Cresap, West Virginia Pilot Plant", J.A. Phinney, Consolidation Coal Co., American Institute of Chemical Engineers, Pittsburgh, June 1974. (13 pages)

31. "Coal Technology : Key to Clean Energy", Office of Coal Research. U.S. Dept. of the Interior, Annual Report, 1973-4 (145 pages)

32. "Coal to Gas System Makes Two Fuels from One Lump of Coal", D.E. Adam, et alia, Garrett Research & Development Company, Coal Mining and Processing, March 1974. (2 pages)

33. "COED Program Achievements", F.M.C. Corp., undated (23 pages)

34. "COGAS Project, The — One Method of Coal-to-Gas Conversion", L.H. Dierdorff & R. Bloom Jr., COGAS Development Co., Soc. Aut. Eng., Portland, Aug. 1973 (6 pages)

35. "Combined Steam Turbine — Gas Turbine Supercharged Cycles Employing Coal Gasification", E.A. Pirsh & W.L. Sage, Sym. on Coal Combustion in Present and Future Power Cycles, Toronto, May 1970. (23 pages)

36. "Combustion Performance of C.O.E.D. Char, The", L.J. Scotti, et alia, F.M.C. Corp., Princeton, New Jersey, January 1975.

37. "Commercial Concept Designs", C.F. Braun & Co., 5th Synthetic Pipeline Gas Sym., Chicago, Oct. 1973 (21 pages)

38. "Conversion of Coal to Gasoline", G. Alex Mills, Industrial and Engineering Chemistry, July 1969. (12 pages)

39. "C.R.G. Methanation", Hydrocarbon Processing, April 1973 (1 page)

40. "C.R.G. Hydrogasification", Hydrocarbon Processing, April 1973 (1 page)

41. "Current Developments in Fuel Utilization", J.C. Taylor, Lurgi (Aust.) Pty. Ltd., Roy, Aust, Chem. Inst., Morwell, July 1974 (34 pages)

42. "Demonstration Plant — Clean Boiler Fuels from Coal: Preliminary design/capital cost estimate ", Vol. 1, Office of Coal Research, U.S. Dept. of the Interior — Ralph M. Parsons Co., Los Angeles, Sept. 1973. (77 pages)

43. "Demonstration Plant — Clean Boiler Fuels from Coal : Preliminary design/capital cost estimate", Vol. 11, Office of Coal Research, U.S. Dept. of the Interior — Ralph M. Parsons Co., Los Angeles, undated. (48 pages)

44. "Dependence of Coal Liquefaction Behaviour on Coal Characteristics", Coal Research Centre, Pennsylvania State University, O.C.R. Report No. 61, June 1974. (40 pages)

45. "Directional Properties of Coal and their Utilization in Underground Gasification Experiments", U.S. Bureau of Mines Technical Report 73, Nov. 1973 (11 pages)

46. "Direct Production of Methane from Heavy Hydrocarbons and Steam, The", N.J. Kertamus & C.D. Woolbert, Babcock and Wilcox Co., 168th Nat. Meeting Am. Chem. Soc. Sept. 1974 (10 pages)

47. "Economic Evaluation and Process Design of a Coal — Oil — Gas (C O G) Refinery", Chem Systems Inc. and Pittsburgh and Midway Coal Mining Co., 65th Annual Meeting, Am. Inst. Chem. Eng. New York, Nov. 1972 (18 pages)

48. "Energy : A Problem to All — A Challenge to LURGI", Lurgi GmbH, Metallges A.G., No. 17, 1974 (7 pages)

49. "Energy Program, 1972", Bureau of Mines, U.S. Dept of the Interior", 1972. (107 pages)

50. "Energy Research Program of the U.S. Dept. of the Interior", Office of Research and Development, Feb. 1974 (draft) 107 pages)

51. "Evaluation of Coal Gasification Technology — Part I — Pipeline Quality Gas", National Academy of Engineering, Washington, 1972 (80 pages)

52. "Evaluation of COED Syncrude", L.J. Scotti, et alia. F.M.C. Corporation, Princeton, New Jersey, January 1975.

53. "Eyes Stay on Coal Conversion", Chemical Engineering, July 1974. (3 pages)

54. "First Commercial Coal to Pipeline Gas Plant is Proposed", H. Levene, Coal Mining and Processing, March 1973 (5 pages)

55. "Fluidized-bed Boiler Project" Electrical Week, Sept. 1973 (1 page)

56. "Fluidized-Bed Boiler to be Built", O.C.R., Dept. of the Interior, News Release, Oct. 14, 1972 (1 page)

57. "Fluidized Bed Holds Promise for Coal", Electric Light and Power, April 1973 (3 pages)

58. "Fluidized-bed Test", Pittsburgh Energy Research Centre, Bureau of Mines, Pittsburgh, March, 1973. (19 pages)

59. "Fuel Gas from Molten Iron Gasification", Applied Technology Corp., Pittsburgh, Pa., September 1973 (18 pages)

60. "Fisher Assay of Oil Shale", The Oil Shale Corp., undated (15 pages)

61. "Gasification and Desulphurization of Coal for Power Generation", S. Lemezis, et alia, Westinghouse Electric Corp., American Power Conference, Chicago, May 1973 (11 pages)

62. "Gas-from-Coal : An Update", Nicholas P. Chopey, Chemical Engineering March 1974. (4 pages)

63. "Gas from Coal with Lurgi Gasification at Sasol", J.C. Hoogendoorn, South African Coal, Oil and Gas Corp. Ltd., Sasolburg, (undated). (14 pages)

64. "Gasification of Coal, The", Harry Perry, Scientific American, March 1974. (7 pages)

65. "Gasification or Liquefaction : Where We Stand", Harold D. Levene, Coal Mining and Processing, Jan. 1974 (6 pages)

66. "Gasoline from Coal via the Synthol Process", L.W. Garrett, Jr., Chemical Engineering Progress, April 1960 (5 pages)

67. "Gas Production", Lurgi, 1971 (brochure 33 pages)

68. "Gasynthan", Hydrocarbon Processing, April 1973 (1 page)

69. "H-Coal : Conversion of Western Coals", C.A. Johnson et alia, Transactions, Soc. Min. Eng., Sept. 1973 (4 pages)

70. "H-Coal Process for Producing Liquid Hydrocarbons", C.A. Johnson et alia, Hydrocarbon Research Inc., 103rd A.I.M.E., Dallas, Feb. 1974. (19 pages)

71. "H-Coal Prototype Plant Program", C.A. Johnson et alia, H.R.I. 66th A.I.Ch.E., Philadelphia Nov. 1973 (22 pages)

72. "High-BTU Gas", O.C.R. Annual Report 1973/74 (50 pages)

73. "High-BTU Gas from Coal — New Mexico Plant Gets Go-ahead", Energy, Pipelines and Systems, Aug. 1974 (3 pages)

74. "HYGAS", Hydrocarbon Processing, April 1973 (1 page)

75. "HYGAS Coal Gasification Process", J. Huebler et alia, Institute of Gas Technology, Chicago, June 1974 (17 pages)

76. "HYGAS Process, The", J. Huebler et alia, Institute of Gas Technology, Chicago, 12th World Gas Conference, Nice, 1973 (20 pages)

77. "I.G.T. HYGAS Process", Institute of Gas Technology, Chicago, undated. (3 pages)

78. "I.G.T. U-Gas [TM] Process", J.W. Loeding and J.G. Patel, Institute of Gas Technology, Chicago, 67th Annual Meeting Am. Inst. Ch. Eng., Washington, Dec. 1974 (13 pages)

79. "Industrial Gas from Sasol", Gas and Coke, Jan 1964 (1 page) (LURGI)

80. "J.G.C. Methane-rich Gas (M.R.G.)", Hydrocarbon Processing April 1973 (1 page)

81. "Kellogg's Coal Gasification Process", A.E. Corer, et alia, Chemical Engineering Progress, March, 1973 (reprint 7 pages)

82. "Keystone Coal Industry Manual, 1974", McGraw-Hill, New York, 1974 (859 pages)

83. "King Coal's Rebirth", Hydrocarbon Processing, May 1971 (3 pages)

84. "Liquid-Phase Methanation — P.D.U. Results", David B. Blum et alia, Chem Systems Inc., 6th Sym. Pipeline Gas Sym., Chicago, Oct. 1974 (23 pages)

85. "Low-BTU Gas from Coal", J.W. Loeding and J.G. Patel, Institute of Gas Technology, Chicago, April 1974 (19 pages)

86. "Low Sulfur Synthetic Crude Oil from Coal", Marvin I. Greene et alia, F.M.C. Corp., Am. Chem. Soc. Los Angeles, April 1974 (18 pages)

87. "LURGI", Hydrocarbon Processing, April 1973 (1 page)

88. "Make Ammonia from Coal", S. Strelzoff, Hydrocarbon Processing, October 1974 (3 pages)

89. "Mathematical Modeling of the Selectivity of the Solvent-Refining Process in the Removal of Sulfur from Coal", R.M. Balwin, et alia, Chemical and Petroleum — Refining Engineering Dept., Colorado School of Mines, 78th Nat. Am. Inst. of Chem. Eng., Golden, Aug. 1974 (44 pages)

90. "METHANOL", Hydrocarbon Processing, April 1973 (1 page)

91. "Methanol from Coal Can Be Competitive with Gasoline", W.D. Harris & R.R. Davison, Texas A. & M. University, The Oil and Gas Journal, Dec. 1973 (3 pages)

92. "Methanol — The New Fuel from Coal", G. Alex Mills and Brian M. Harney, U.S. Bureau of Mines, Chemtech, Jan 1974 (6 pages)

93. "Methanation of Coal for SNG", F.W. Maeller, et alia, Hydrocarbon Processing, April 1974 (6 pages)

94. "Minerals Research in C.S.I.R.O.", No. 6 C.S.I.R.O. Minerals Research Laboratories, North Ryde June 1974 (11 pages)

95. "Mining Oil from Coal", Mining News in Australia, Nov. 1974 (3 pages)

96. "Multi-Stage Fluidized-Bed Pyrolysis of Coal at the Project COED Pilot Plant", Louis J. Scotti et alia, F.M.C. Corp., 77th Nat. Am. Inst. Chem. Eng. Pittsburgh, June 1974 (31 pages)

97. "New Coal Gasification Technology, The", Frank C. Schora Jr., et alia, Institute of Gas Technology, Chicago, Annual A.I.M.E., Dallas, Feb. 1974. (26 pages)

98. "New Concepts in High-BTU Technology", R. Detman, C.F. Braun and Co., Coal Processing Technology, New York, 1974 (5 pages)

99. "New Processes Brighten Prospects of Synthetic Fuels from Coal", Keystone Coal Industry Manual, 1974 (Reprint) (10 pages)

100. "New Technology for Conversion of Coal to Pipeline Gas", R. Detman, C.F. Braun & Co., Soc. Pet. Eng., A.I.M.E. Indianapolis, March 1974. (20 pages)

101. "North Dakota : Proposed Coal Gasification Complex will Tap Lignite Reserves", Coal Age, May, 1974 (2 pages)

102. "Observations of Coal Processing in an Electrofluid Reactor", A.H. Pulsifer & T.D. Wheelock, Dept. of Chem. Eng., Iowa State University, Ames, undated (30 pages)

103. "O.C.R. to Build Pollution-free Test Boiler at Powerplant Site in Rivesville, West Virginia", O.C.R. Dept. of the Interior, News release, July 3rd, 1973 (5 pages)

104. "Oil and Gas from Coal — A review of the Current International Situation as it Applies to Queensland", Dept. of Commercial and Industrial Development, Brisbane, March 1974. (20 pages)

105. "Oil Shale", TOSCO, undated (5 pages)

106. "Oil Shale — A Solution to the Road-block", C.W. Livingston, Barodynamics Inc., Engineering and Mining Journal, June 1974 (9 pages)

107. "Opportunities for Ammonia Production via Coal Gasification, The," Stephen N. Rothman and Marshall E. Frank, Chem. Systems Inc., 78th A.I.Ch.E., Salt Lake City, Aug.1974 (8 pages)

108. "Plants for the Gasification of all Kinds of Fuels by Koppers-Totzek Process", Heinrich Koppers GmbH, Essen, June 1974 (2 pages)

109. "Pilot Plant for De-ashed Coal Production", V.L. Brant and B.K. Schmid, Stearns-Roger Corp., Denver, Chemical Engineering Progress, Dec. 1969 (6 pages)

110. "Potential for Synthetic Oil", Petroleum Press Service, July 1973 (3 pages)

111, "Preliminary Results Released for Wyoming 'In-Situ' Gasification Test", L.A. Schrider and J. Pasini, U.S. Bureau of Mines, Coal Age, December 1973 (4 pages)

112. "Present Status of the H-Coal Process", C.A. Johnson et alia, H.R.I., Trenton, Sept. 1973 (27 pages)

113. "Prepare Coal for Gasification", J.H. Dressel and Joseph Absil, Hydrocarbon Processing, March 1974 (3 pages)

114. "Process Industries in Australia — Impact and Growth", 2nd Nat. Chem. Eng. Conference, Surfers Paradise, July 1974 (505 pages)

115. "Process to Make High-B.T.U. Gas from Coal, A", Albert J. Forney et alia, U.S. Bureau of Mines, Technical Progress Report 24, Pittsburgh, April 1970 (6 pages)

116. "Processing Coal to Produce Synthetic Crude Oil and a Clean Fuel Gas", Haig D. Terzian et alia, F.M.C. Corp., Princeton, Am. Inst. Ch. Eng., Los Angeles, April 1974. (20 pages)

117. "Production of Gasoline from Australian Brown Coal by the H-Coal Process", C.A. Johnson et alia, H.R.I. Trenton, 8th World Petroleum Congress, undated. (8 pages)

118. "Production of Liquid Fuel from Coal by the Lurgi-Ruhrgas/Tar Hydrogenation Process, The", Lurgi, undated (13 pages)

119. "Production of Reduction Gas from Coal and Fine Coke for the Iron and Steel Industry", H. Staege, Heinrich Koppers GmbH, Essen, May 1974 (39 pages)

120. "Production of Liquid Fuels from Coal, Developments 1972-74, The", D.J. Nicklin, Dept. of Chem. Eng. University of Qld., Feb. 1975.

121. "Recent Developments in Liquid Phase Methanation", Martin B. Sherwin et alia, Chem Systems Inc. 5th Synthetic Pipeline Gas Sym., Chicago, Oct. 1973 (25 pages)

122. "Removal of Sulfur from Coal by Treatment with Hydrogen", J.H. Gary et alia, Chemical Eng. Dept., Colorado School of Mines, Golden, 167th A.C.S., Los Angeles, March 1974 (24 pages)

123. "Research, '72" Bureau of Mines, U.S. Dept. of the Interior, Washington, 1972 (82 pages)

124. "Retorting of Coal, Oil Shale and Tar Sand by Means of Circulated, Fine-Grained Heat Carriers as a Preliminary Stage in the Production of Synthetic Crude Oil, The", Ronald W. Rammler, Lurgi, Frankfurt, A.I.M.E. Annual Meeting, Denver, Feb. 1970 (38 pages)

125. "Review of Methanation Demonstration at Westfield, Scotland", James E. Landers, Conoco Coal Development Co., 1974 (5 pages)

126. "Solution Hydrogenation of Lignite in Coal and Derived Solvents", D.E. Severson, et alia, University of North Dakota, Grand Forks, North Dakota, 1968 (19 pages)

127. "Solvent Extraction of Coal — Batch Autoclave Studies", A.C.I.R.L., Aug. 1967. (81 pages)

128. "Solvent Extraction of Coal — Further Laboratory Studies", A.C.I.R.L., Nov. 1969 (20 pages)

129. "Solvent Refined Coal", Pittsburgh and Midway Coal Mining Co. — Gulf Oil Co., (undated brochure)

130. "Solvent Refined Coal Process, The", B.K. Schmid, The Pittsburgh and Midway Coal Mining Co., Ft. Lewis Sym. on Coal Gasification and Liquefaction, Pittsburgh, Aug. 1974. (10 pages)

131. "S.N.G. from Coal via the Lurgi Gasification Process", James M. Moe, Fluor Engineers and Constructors Inc., Los Angeles, undated (19 pages)

132. "S.N.G. Plans Shift to Coal", ed., The Oil and Gas Journal, Aug. 1974. (7 pages)

133. S.N.G. Production Based on Koppers-Totzek Coal Gasification", J.E. Franzen and E.K. Goeke, Heinrich Koppers GmbH, Essen, 6th Synthetic Pipeline Gas Symposium, Chicago, October 1974, (20 pages)

134. "Status of the COGAS Project", Ralph Bloom Jr., and R. Tracy Eddinger, COGAS Development Co., Princeton, 6th A.G.A. Synthetic Pipeline Gas Sym., Chicago, Oct. 1974 (22 pages)

135. "Status of the HYGAS Program", B.S. Lee and P.B. Terman, Institute of Gas Technology, Chicago, 6th A.G.A. Synthetic Pipeline Gas Sym., Chicago, Oct. 1974 (20 pages)

136. "Status of Coal Liquefaction and Gasification Technology", W.R. Epperly, Esso Research and Engineering Co., Baytown, Guilford Centre Eng. Sym., Greensboro, Feb. 1974 (21 pages)

137. "Status of the O.C.R./A.G.A. Coal Gasification Program", Roger Detman, C.F. Braun & Co., 3rd A.G.A. – I.G.T. Conf. on Nat. Gas Research & Technology, Dallas, March 1974. (13 pages)

138. "Status of the Steam-Iron Process", P.B. Tarman, Institute of Gas Technology, Chicago, 5th A.G.A. Synthetic Pipeline Gas Sym., Chicago, Oct. 1973 (11 pages)

139. "Status of the Technology – Coal Gasification, Oil from Coal, The", National Coal Research Advisory Committee, Report to The Minister for Minerals and Energy, Melbourne, May 1973 (25 pages)

140. "Status Report : The A.G.A./O.C.R. Coal Gasification Program", ed. Coal Age, Jan. 1973 (6 pages)

141. "Substitute Natural Gas from Coal", C.G. Van Fredersdorff and F.E. Vanderreer (undated - 23 pages)

142. "SYNTHANE", Hydrocarbon Processing, April 1973 (1 page)

143. "Synthane Prototype Pilot Plant", U.S. Bureau of Mines, Bruceton, undated. (1 page)

144. "Synthetic Fuel Activities in the Western Hemisphere", L.E. Swabb, Jr., Exxon Research and Engineering (U.S.A.), German Soc. for Min. Oil Sc. and Coal Chem., Annual Conv., Hamburg, Sept. 1974 (21 pages).

145. "Synthesis Gas Through Gasification of All Kinds of Fuel by the Koppers-Totzek Process", Heinrich Koppers Gmbh, Essen, July 1973 (6 pages).

146. "Technology and Cost of Coal Gasification", H.M. Siegel and T. Kalina, Esso Research and Engineering Co., Flotham Park, Mechanical Engineering, May 1973. (6 pages).

147. "They're Making a Solid Effort to Get Clean Coal Liquids", D.P. Burke, Chemical Week, September 1974. (7 pages).

148. "TOSCOAL Process, The – Coal Liquefaction and Char Production", Franklin B. Carlson et alia, The Oil Shale Corp., Golden, Am. I. Ch.E., Pittsburgh, June 1974. (28 pages).

149. "Upgrade Low Btu Gas", J.K. Jones, Petrocarbon Developments, Hydrocarbon Processing, September 1973. (3 pages).

150. "Upgrading of Solid Fuels", Lurgi, Frankfurt, March 1971. (Brochure) (27 pages).

151. "Utilization of Queensland Coals", J. Job and D.J. Nicklin, Dept. of Chemical Engineering, University of Queensland Dec. 1967 (97 pages).

152. "Utilization of Queensland Coals, The –(Developments 1968 – 1972)", D.J. Nicklin et alia, Dept. of Chem. Eng. Uni. of Qld., Brisbane (undated). (79 pages).

153. "United States Steel Research Laboratory", Coal Age, Oct. 1973 (5 pages).

154. "What's Ahead for S.N.G. from Liquids and Coal?", G. Long, Northern Illinois Gas Co., Pipe Line Industry, Oct. 1973. (3 pages).

155. "When Will Synthetic Fuels Begin Flowing from Coal?", Chemical Week, Jan. 1974. (2 pages).

APPENDIX I

Major Governmental, Semi-Governmental and Private Agencies sponsoring Coal Conversion Research Programs

ALBERTA PROVINCIAL GOVERNMENT OF CANADA

The Provincial Government is considering development of "in situ" coal gasification projects. A feasibility study is currently underway, and the neighbouring Province of British Columbia has expressed interest in this study.

The Research Council of Alberta promotes research into coal conversion and is the holder of a patent which describes a process for the separation of extracts from coal-oil slurry.

AMERICAN GAS ASSOCIATION (A.G.A.)

The A.G.A. is an association of companies involved in the U.S. gas industry which is undertaking a research and development program aimed at producing substitute natural gas from coal. As part of the A.G.A.'s long-term "Gas Industry Research Plan, 1974-2000", the S.N.G. research program includes work on:

1. The production of S.N.G. from coal, naphtha, and organic wastes.
2. Coal and oil shale mining.
3. Construction materials in S.N.G. plants.
4. The production of hydrogen from water.
5. Experimental effects in S.N.G. plants.

Annual expenditures are recommended for the 1974-78 period within each of these categories. The estimated cost of the S.N.G. research and development program for that period is about U.S.\$1,750 million, which amounts to approximately 75% of the entire gas industry's research and development program for the same period. Total cost of all S.N.G. programs is estimated at U.S. \$7,200 million.

A.G.A. and O.C.R. have entered into an Agreement to co-sponsor the development of high-BTU coal gasification processes. Details of the Agreement are given in the section on the Office of Coal Research.

BATTELLE MEMORIAL INSTITUTE

The Battelle Memorial Institute conducts research and development at laboratories in Columbus, Ohio, and at the Pacific Northwest Laboratories in Richland, Washington.

Battelle has established the "Battelle Energy Program" to carry out "major R & D efforts aimed at developing practical solutions to some of the extremely serious energy-shortage and utilisation problems expected to exist in the United States during the 1975-1995 time period".

Battelle Memorial Institute has allocated U.S.\$25 million of its own funds to support the 5-year Energy Program in which the primary emphasis will be placed on coal mining and conversion to clean fuels and chemicals.

With Union Carbide Corp., Battelle is developing the AGGLOMERATING ASH gasification process for the production of med-and high-BTU gas.

Note: Battelle also has laboratories and other facilities as follows:

Battelle Institute, V., Frankfurt, W. Germany.
Geneva Research Centre, Geneva, Switzerland.
Corporate Offices are located at 505 King Avenue, Columbus, Ohio.

BRITISH GAS COUNCIL

The British Gas Council, previously The Gas Council, sponsors the work at the Midlands Research Station on high-pressure fluidised-bed models, fluidised-bed gasification, hydrogenation, and methods of upgrading coal-derived gas to pipeline quality.

In its own right the British Gas Council holds various patents relating to coal conversion.

COMMONWEALTH SCIENTIFIC AND INDUSTRIAL RESEARCH ORGANISATION (C.S.I.R.O.)

The utilisation of Australian coals has been studied by C.S.I.R.O. since 1949, first by a Coal Research Section and now by the divisions in the Minerals Research Laboratories group.

The major effort of C.S.I.R.O.'s program has been directed towards assessing coals in terms of their physical and chemical properties and relating these properties to existing and possible utilisation processes. Studies have been made on various coal-conversion techniques including solvent extraction, combustion, gasification, carbonisation and the properties and upgrading of coal tars.

C.S.I.R.O. has placed considerable emphasis on the study of carbonisation of Australian coals, utilising fixed-bed and fluidised-bed techniques to determine the effect of coal type, temperature and additives on the yields and properties of tar, oil, gas and coke. Investigations into the cracking of tars and other related materials in fixed- and fluidised-bed units to produce simpler aromatic hydrocarbons, pitch, gas and electrode carbon form part of the program.

An extensive study on the hydro-gasification of brown and bituminous coals to produce methane-rich gas for fuel use has also been undertaken. A technical-scale study on fluidised-bed gasification of coal with air and steam to produce a gas rich in hydrogen and carbon monoxide has been developed. A combined moving-burden and entrained-flow process was devised during the fluidised-bed gasification project.

Current studies on the gasification of chars with steam to produce synthesis gas show promise.

ELECTRIC POWER RESEARCH INSTITUTE (E.P.R.I.)

The Electric Power is a non-profit research institute, formed in 1972, to expand electrical energy research and development under the sponsorship of the U.S. utility industry.

The E.P.R.I. sponsors research and development of processes that promise to provide suitable fuels from coal for power generation.

A demonstration plant based on Lurgi coal gasification is under construction at Commonwealth Edison Co's Powerton station near Pekin, Ill. under the joint financing of Commonwealth Edison and E.P.R.I. (See "Lurgi" in Section III for further details).

Another project in which E.P.R.I. is interested, is the construction of a 120 T.P.D. pilot plant at Windsor, Connecticut, based on a coal gasification process being developed by Combustion-Engineering Inc. for combined-cycle power generation.
(See "C-E Entrained Fuel Process" in Section III).

The 1975 appropriation by E.P.R.I. for research and development is U.S. $108 million of which $27 million is for fossil fuels research.

E.P.R.I. and O.C.R. are working together to develop advanced power generation techniques including combined cycle and M.H.D. generation units.

Within E.P.R.I., the Coal Utilisation Task Force has the responsibility for coal research programs. The Task Force has recommended that E.P.R.I. co-sponsor with O.C.R. and an industry team, the construction and operation of a demonstration plant producing 2,600 barrels per day of low-sulphur distillate fuel or an equivalent of desulphurised solid fuel (S.R.C.) for electric utilities. The governor of Ohio has requested Federal funds of U.S. $43 million through O.C.R. to support this project.

Recently, E.P.R.I. awarded a U.S. $375,000 contract to Pennsylvania State University for a 3-year study

86

on coal liquefaction reaction mechanisms and on solvent-refined coal composition. A further recent contract awarded by E.P.R.I. was to Mobil Research and Development Corp. for research aimed at indentifying problems in utilising raw and processed coal liquids as turbine fuels. Under the terms of this latter contract, E.P.R.I. will contribute 75% of the research funds to be expended.

ENVIRONMENTAL PROTECTION AGENCY (E.P.A.)

The E.P.A. funds research and development on the utilisation of U.S. coal resources in an environmentally-acceptable manner. To this end, E.P.A. has undertaken a joint study with the U.S. Dept. of Health, Education and Welfare on the effects of sulphur emissions on public health. The results of this study prompted E.P.A. to initiate sponsorship of R & D on coal conversion processes for producing clean fuels from coal.

Contracts totalling $1,859,350 have been awarded by E.P.A. to two subsidiaries of International Systems and Controls Corp. which call for development of a power-generating unit utilising the Applied Technology Corp's. Two-Stage (Submerged Coal) process to process high-sulphur coals within E.P.A. pollution standards. Applied Technology will develop design criteria for a 50- to 100- MW power generating unit utilising the Two-Stage gasification process (see Section III) under a $1,719,350 contract. J.F. Pritchard & Co., the other subsidiary of ISC, received a $140,000 contract for an engineering evaluation of the Two-Stage Process. This contract also calls for a preliminary design of a power generating plant with a molten-iron gasifier coupled with a combined-cycle power generating unit.

INDUSTRIAL DEVELOPMENT CORPORATION OF RHODESIA (I.D.C.)

In March 1974, it was announced that the Ministry of Mines, under the direction of I.D.C., would undertake, in conjunction with the Institute of Mining Research at the University of Rhodesia, a full-scale investigation of the production of oil from coal.

INSTITUTE OF GAS TECHNOLOGY (I.G.T.)

The Institute of Gas Technology is an affiliate of the Chicago Institute of Technology, and is primarily engaged in research and development of coal gasification processes for the production of low-, med.-, and high- BTU gas within the United States.

The HYGAS process is under development by I.G.T. with the support of the American Gas Association and the Office of Coal Research. I.G.T. also offers a wide range of support services to the gas industry.

INTERNATIONAL COMMITTEE FOR COAL RESEARCH (I.C.C.R.)

The committee was formed in 1972 to encourage the continuous exchange of information and experience in coal research between coal producers and research institutes in the member countries.

Members of the I.C.C.R. represent the coal industries of Western Europe, the United States, Canada, South Africa and Australia.

MINISTRY OF INTERNATIONAL TRADE AND INDUSTRY OF JAPAN (MITI)

In November, 1973, an agreement, in principle, was entered into between the then Minister of MITI, Mr. Nakasone, and the Australian Minister for Minerals and Energy, Mr. R.F.X. Conner, relating to the joint examination on the feasibility of establishing coal-conversion plants in Australia. Up to the present time (May 1975) no firm agreement has been ratified.

The Agency of Industry and Technology, which is a department of MITI, has been involved for some years in the conversion of coal to liquid and gaseous products.

During mid-1974 results were released on their hydrogenation process, "SOLVOLYSIS", and detailed experimentation and research on this process and others, is continuing.

NATIONAL COAL ASSOCIATION (N.C.A.)

The U.S. coal industry, through membership in the National Coal Association, finances coal research by the N.C.A. research affiliate, Bituminous Coal Research, Inc. (B.C.R.) of Pittsburgh, Pa.

Bituminous Coal Research was chartered as a corporation in 1933 and its purpose is "exclusively to encourage, foster and promote in any lawful manner the utilisation of coal and its products by means of study, research and education".

In 1956, B.C.R. established a chemically-oriented laboratory in Pittsburgh. In October 1961, research operations were consolidated in a major research centre in Monroeville where, in addition to its "in-house" projects, B.C.R. has conducted research for such groups as the Office of Coal Research, U.S. Bureau of Mines, the U.S. Environmental Protection Agency, the Commonwealth of Pennsylvania and the electric utility industry.

B.C.R. is the developer of the BI—GAS gasification process and conducts research in all facets of coal conversion technology.

NATIONAL COAL BOARD OF THE UNITED KINGDOM

In the United Kingdom, the National Coal Board conducts research and development on mining and coal conversion through sponsorship of the Coal Research Establishment (C.R.E.) at Stoke Orchard.

A process under investigation at C.R.E. promises to open the way to augmenting coking technique by-products by commodities made directly from coal. The process is of the S.R.C. type, employing the digestion of the coal in a hydrogen-donor solvent (e.g. anthracene oil) with subsequent filtration to remove insoluble residue from the coal solution. The preparation of electrode coke and carbon fibre by this route is under study and processes for exploiting the work commercially are being developed.

In an experimental plant at C.R.E., coal can be continuously digested at a rate of 15 kg/hour and the digest filtered and either converted into S.R.C. by evaporation of the solvent, or coked.

It is believed by C.R.E. that hydrogenation of coal in solution would probably be the first step in the development of coal liquefaction processes and it has been shown that by catalytic hydrogenation of the coal extract, coal can be converted to a wide range of liquid products.

Another process being developed at Stoke Orchard is gas extraction of coal. In gas extraction, coal is treated with compressed gases at temperatures of 350-400°C, causing a portion of the coal to go into solution in the compressed gas and the ash is left with unreacted coal. The resultant solution is transferred to a vessel at lower pressure where the gas is separated for recycling leaving a refined coal extract having constituents with lower molecular weights than those obtained by solvent extraction. However, yields of extracts with compressed gases are lower than with liquid solvents but can be increased significantly by utilising gas extraction in the presence of reducing agents such as hydrogen, or carbon monoxide and steam. The gas extraction residue is a porous solid suitable for gasification. As no filtration step is required, and recovery of gaseous solvent gases is virtually complete, the process offers some attractive features.

There are experimental gas extraction units at Stoke Orchard evaluating the preparation and recovery of coal extract to provide design data for a pilot plant.

NATIONAL COAL RESEARCH ADVISORY COMMITTEE OF AUSTRALIA.(N.C.R.A.C.)

The committee is comprised of representatives of the mining industry, gas and electricity producers, C.S.I.R.O. and the Dept. of Minerals and Energy.

In May, 1973, N.C.R.A.C. presented a report to the Minister for Minerals and Energy stating that, in its opinion no substantial Federal government funds should be expended on research and development of coal-conversion processes, however, a watching brief should be kept on all overseas technological advances.

In April 1975, the committee recommended three major avenues worthy of detailed evaluation. There were:

1. The definition, by 1980, of the nation's coal resources with particular attention to opencut coal suitable for the manufacture of coke and the production of synthetic hydrocarbon fuels.

2. A research program aimed at providing, by 1985, the necessary background data for the development of a rational program for mining, transporting and the utilisation of Australian coal.

3. The detailed and comprehensive assessment of overseas and indigenous coal-conversion technology and the development of coal conversion processes to provide, by 1990, acceptable substitutes for all imported hydrocarbon fuels.

The Australian Coal Industry Research Laboratories (A.C.I.R.L.) is sponsored by N.C.R.A.C. in all phases of the utilisation of Australian coals.

Note: Individual Australian States are also examining the utilisation of State coal reserves as follows:

QUEENSLAND.

The Queensland Government in 1946 commissioned a study by the Powell-Duffryn Technical Services Group into Queensland's coal industry, with the main report being published in 1949. A supplementary study on the conversion of coal into oil was presented to the government in 1952. In the light of current events, it is interesting to note that this latter report emphasised the potential importance of the Walloon Coal Measures in south-east Queensland for conversion. However, the recommendations of this report were not immediately implemented and it was not until 1969 that private enterprise commenced detailed exploration in the area. The potential of the region is emphasised by announcements in late 1974 by the Mitsui Group of Japan concerning the Millmerran Coal company's areas inland from Brisbane (see S.R.C.–MITSUI and S.R.C.–PAMCO for further details). Various statements by the Federal Minister for Minerals and Energy also highlight this area.

Another area of potential importance to the State Government is the Galilee Basin of central-western Queensland outlined by scout drilling of the Mines Dept. drilling branch during 1973-74 and subsequently offered for tender (in part) to companies interested in coal conversion. Despite various optimistic press reports this area is believed to afford little promise for the delineation of significant open-cut reserves suitable for economic conversion to liquid products. The reserves within this basin could eventually prove suitable for degasification or "in situ" gasification.

Since 1967, the Department of Commercial and Industrial Development has sponsored various studies by the Dept. of Chemical Engineering of the University of Queensland, St. Lucia, into the utilisation of Queensland coals with the most recent report being released in February, 1975, covering the period 1972-1974. Studies are also in progress at the Queensland Institute of Technology on batch-autoclave experiments in the solvent refining of coal, and the examination of hydrogenation catalysts.

The Queensland National Party (State Coalition majority party) formed, in December 1973, a party Committee on Energy Resources to evaluate and review matters relating to energy on the party's behalf. The State Energy Resources Advisory Council, made up of Government officers, conducts a similar exercise, reporting to the Minister for Mines and Energy, who, in turn, reports to the State Cabinet.

NEW SOUTH WALES

Whilst, to date, no body has been established for the specific purpose of utilising the State's coal reserves for conversion, the State Government has expressed interest in this field and officers of the N.S.W. Mines Department are monitoring overseas research in conjunction with the evaluation of the State's coal reserves for suitability for conversion to liquid and gaseous products.

SOUTH AUSTRALIA

In late 1974 the State Government established the South Australian Energy Committee to examine the possibility of utilising the State's coal deposits for the production of liquid fuels.

The two areas of greatest interest are at Lake Phillipson in the north of the State, held by Utah Development Company and in the Inkerman-Balaklava area in the mid-north, presently under investigation by the Electricity Trust of South Australia.

A recent A.C.I.R.L. study was commissioned on the utilisation of these deposits and both the Energy Committee and the State Dept. of Mines maintain a close liaison with the C.S.I.R.O., A.C.I.R.L., and N.C.R.A.C.

VICTORIA

During 1973, the Victorian Government opened discussions with the German Democratic Republic relative to the State's vast (estimated at 85,000 million tonnes) reserves of brown coal in the La Trobe Valley approximately 80 miles south-east of Melbourne. In December 1973, the State Premier officially announced that his government had accepted a proposal from the G.D.R. to cooperate on brown coal-to-oil technology. At the same time it was announced that the G.D.R. would contribute $100,000 towards the establishment of a special department within a Victorian university to provide facilities for the study of brown-coal technology.

An agreement between the Victorian Government and the G.D.R. was announced in mid-April 1975, which allowed for the joint development of coal-conversion technology specific to the conversion of brown coal to petrol.

Private research carried out since the early 1970's by Morwell Industrial Development Pty. Ltd. and International Oil Ltd. has concentrated on the conversion of brown coal reserves held by those companies. In late 1973, Morwell Industrial Development Pty. Ltd. announced plans to construct a small oil refinery and methanol plant utilising a mix of 73% Bass Strait crude and 27% oil from coal. Early 1974, International Oil Ltd. announced the results of talks with a number of overseas companies involved in coal conversion relative to the establishment of a A$250-300 million facility to convert coal to oil.(based on either the S.R.C. process or the H-COAL process). By mid-1974 an apparent change in the company's policy was indicated following an announcement that the company could, under certain conditions, take up the rights to the C.S.I.R.O. brown coal demineralisation process which is suitable for the production of electrode and active carbon. Concurrently, the company is examining the production of liquid fuels from lignite.

WESTERN AUSTRALIA

The State Government commissioned a report on the "in situ" gasification of deep coal seams known to exist about 150 miles north of Perth. The coal seams occur over an area of about 1,750 square miles and inferred reserves are of the order of 35,000 million tons, however these seams are inaccessible and could not be economically extracted by present methods of mining.

The report predicts that these deep seams will be exploited in the future by "in situ" gasification. (or degasification techniques) The Fuel and Power Commission is examining all aspects of the States future energy needs.

In August 1974 a further study was commissioned to examine the available techniques and costs associated with the conversion of "Collie" Coal to liquid and gaseous products.

NATIONAL RESEARCH DEVELOPMENT CORPORATION (N.R.D.C.)

In the United Kingdom, N.R.D.C. sponsors work being carried out by the British Coal Utilisation Research Association Ltd. (B.C.U.R.A.) at Leatherhead on coal utilisation, primarily fluidised-bed combustion and the production of formed-coke, which have some application to the problem of coal conversion.

NATIONAL SCIENCE FOUNDATION (N.S.F.)

A wide range of U.S. research work is sponsored by the National Science Foundation through its RANN Division.

N.S.F.–RANN sponsors experimental studies on coal conversion at the Pennsylvania State University.

OFFICE OF COAL RESEARCH

(From Jan. 1975 within the Energy Research & Development Administration)

The Office of Coal Research (O.C.R.) was established within the U.S. Department of the Interior on 7th July, 1960 under an Act of Congress.

The duties of O.C.R. as outlined in the Act are to:

1. Develop, through research, more efficient methods of mining, preparing and utilizing coal.

2. Contract for sponsor, co-sponsor, and promote the coordination of, research with recognised interested groups including, but not limited to, coal trade associations, coal research associations, educational institutions, and Agencies of States and political sub-divisions of States.

3. Establish technical advisory committees composed of recognised experts in various aspects of coal research to assist in the examination and evaluation of research progress and of all research proposals and contracts, and to ensure the avoidance of duplication of research.

4. Co-operate to the maximum extent with other departments, agencies and independent establishments of the Federal and State Governments, and with all other interested organisations, governmental and non-governmental.

Since its establishment, O.C.R. has assisted the development of many significant coal conversion projects. A full listing of O.C.R.-sponsored projects can be obtained from the main body of this report.

On 3rd August, 1971, an Agreement was entered into between O.C.R. and the American Gas Association (A.G.A.) to provide for the joint funding and acceleration of the coal gasification program.

Under this Agreement, A.G.A. would contribute up to U.S. $10 million per annum for four years commencing in fiscal-year 1972, to match, on a two-thirds to one-third ratio, Federal funding appropriated by Congress through O.C.R. On this basis, total joint funding of U.S. $30 million annually would be provided during the four-year period. The projects co-sponsored by A.G.A. and O.C.R. under this Agreement are revealed in Section II of this report.

In January, 1975, O.C.R. was removed from the U.S. Dept. of the Interior and incorporated within a new agency, the Energy Research and Development Administration, which was established to unite all Federal energy research and development programs including solar, geothermal, nuclear fission and fusion, and coal.

E.R.D.A.'s fiscal 1976 request for R & D in coal liquefaction and gasification is U.S. $206 million, out of a total request for U.S. $323 million. Appropriations to the Office of Coal Research since its inception were:

Fiscal Year	Appropriations
1961	U.S. $1,000,000
1962	1,000,000
1963	3,450,000
1964	5,075,000
1965	6,836,000
1966	7,220,000
1967	8,220,000
1968	10,980,000
1969	13,700,000
1970	15,300,000
1971	17,160,000
1972	30,650,000
1973	43,490,000
1974	123,400,000
1975	327,330,000

U.S. BUREAU OF MINES (U.S.B.M.)

The Bureau of Mines, of the U.S. Department of the Interior, conducts coal conversion research at various research stations located throughout the United States.

The Bureau is currently developing the "Hydrane" and "Synthane" processes for the production of S.N.G. from coal, the "Stirred Fixed-Bed" process ("MORGAS") for low-BTU gas production, and the "Synthoil" and "Costeam" processes for the conversion of coal to liquid products.

The fiscal 1975 budget of the Bureau is U.S. $212,347,000 of which U.S. $57,022,000 is for coal conversion research and development. The fiscal 1974 budget was U.S. $105,606,000 and the large increase in funds for 1975 indicates the Bureau's increasing emphasis on research and development in coal conversion.

Note: In an announcement in April, 1975, it was reported that the Bureau of Mines had lost all its research programs on coal conversion, coal combustion and petroleum extraction to the newly-created Energy Research and Development Administration (E.R.D.A.).

RESEARCH AND DEVELOPMENT IN OTHER COUNTRIES (not covered in text)

1. **Federal German Republic**

    (a) Steinkohlenbergbauverein —A technical-scientific organization, engaged in coal utilisation along with other private and semi-governmental committees.

    (b) Deutscher-Braunkohlen-Industrie Verein (DEBRIV) is a joint organisation in which all members and organisations of the lignite industry participate. The group carries out its own research into the utilisation of lignite and also contracts outside research bodies.

2. **German Democratic Republic (G.D.R.)**

The G.D.R. is the world's largest miner of lignite with an annual production of about 260 million tonnes. A significant proportion of this lignite is processed into briquettes for utilisation in steam-oxygen gasification for town gas production and Fischer—Tropsch synthesis to a wide range of liquid fuels and chemical by-products. Another significant proportion of lignite production is carbonised at low temperatures to yield tars which are subsequently catalytically hydrogenated to paraffins and lubricating oils, or converted to electrode carbon by coking.

3. **India**

The Geological Survey of India plans to give top priority to the location of new coal reserves to help reduce the country's dependence on oil.

The government has sponsored research and development on hydrogenation of half coal half crude oil mixtures leading to a proposal for consideration by the government for a commercial plant based on the half and half process.

Three coal-based fertilizer plants have been approved by the Indian Government. The first of these is being constructed at Korba under the management of the Fertilizer Corporation of India. This plant will have an annual capacity to produce 495,000 tons of urea.

Other projects under evaluation by the Government include proposals from Lurgi GmbH for the establishment of small (1 million tpy feed) liquefaction plants, and also an SRC plant, considered by Singerani Co., of 3 million tpy feed.

APPENDIX II

ALPHABETICAL LISTING OF COMPANIES
HOLDING MAJOR COAL CONVERSION PATENTS.

Allied Chemical Corporation.
2,977,299; *"Carbonisation Chemical Products"*

Air Products and Chemicals Inc.
3,779,725; *"Coal Gasification".*

Ashland Oil and Refining Company.
3,533,938; *"Blended Coal Products".*

Atlantic Richfield Company.
3,503,867; *"Synthetic Crudè"*
3,619,404; *"Asphaltene Formation during Hydrogenation".*
3,730,694; *"Fuel Gas Upgrading".*

Atomic Energy Commission (U.S.)
3,794,116; *"In-Situ Coal Bed Gasification".*

Bennett Engineering Company.
3,576,734; *"Low-Temperature Carbonisation".*

Bluemner E.
2,714,086; *"Chemical Modification".*

British Gas Council, The
3,390,971; *"Gasification of Solid Carbonaceous Fuel".*
3,607,158; "Fluidised-Bed Coal Hydrogenation Producing Methane Rich Gas".

Chevron Research Company.
3,518,182; *"Motor Fuels via Hydro-visbreaking".*
3,759,677; *"Catalytic Synthesis Gas Manufacture".*
3,775,072; *"Catalytically Steam-Reforming Organic Gasification".*

Cities Service Oil Company.
3,734,180; *"In-Situ Gasification-with Non Hypersensitive Explosives".*
3,734,184; *"In-Situ Coal Gasification".*
3,770,398; *"In-Situ Coal Gasification Process".*
3,775,073; *"In-Situ Gasification of Coal by Seam Fracturing".*

Compagnie Francaise des Essences Synthetiques Societe Anon. France.
2,707,163; *"Solvent Extraction and Fractional Distillation".*

Consolidation Coal Company.
3,018,241; *"Hydrogen-Rich Liquids".*
3,018,242; *"Gasoline via Extraction".*
3,117,921; *"Hydrogen-Rich Liquid Conversion".*
3,120,474; *"Conversion".*
3,143,489; *"Solvent Extraction".*
3,158,561; *"Freidel-Crafts Catalysts Liquefaction".*
3,162,594; *"Hydrogenation of Solvent Extracts".*
3,184,401; *"De-Ashing".*
3,232,861; *"De-ashing Acids".*
3,355,376; *"Zinc Halide Polynuclear Aromatics"* (Jointly with U.S. Secretary of the Interior).
3,516,808; *"Carbon Dioxide Acceptors — Melt Process"* (Jointly with U.S. Secretary of the Interior).
3,523,886; *"Solvent Extraction"* (Jointly with U.S. Secretary of the Interior).
3,692,505; *"Fixed Bed Coal Gasification"*
3,700,422; *"Steam-Iron Process".*

Esso Research and Engineering Company.
  2,741,549; *"Conversion to Volatile Products".*
  3,488,279; *"Two-Stage Conversion"*
  3,488,280; *"Catalytic Hydrogenation – Water Recycle".*
  3,514,394; *"Cyclic Process"*
  3,575,847; *"Sperical Catalysts".*
  3,733,186; *"Two-Step Controlled Flow Gasification Process".*

F.M.C. Corporation.
  3,453,202; *"Hydrogenation – Iodine".*
  3,574,065; *"Fractional Carbonisation".*

Fossil Fuels Inc.
  3,030,297; *"Rapid Hydrogenation".*

Great Lakes Carbon Corporation.
  3,379,638; *"Ash-Free Anode Carbon*(Jointly with The Lummus Co.)

Gulf Research and Development Company.
  2,654,695; *"Hydro-Coking".*
  3,341,447; *"Solvent Processing"* (Jointly with U.S. Secretary of the Interior).

Huntington Chemicals Corporation.
  3,107,985; *"Continuous Distillation – Autogenous Hydrogenation".*

Hydrocarbon Research Inc.
  2,885,337; *"Partial Hydrogenation".*
  25,770 (reissue); *"Gas Liquid Contracting Process".*
  3,321,393; *"Ebullating Bed Catalytic Hydrogenation".*
  3,519,553; *"Ebullating Bed Catalytic Hydrogenation".*
  3,519,554; *"Residium Recovery".*
  3,519,555; *"Ebullating Bed Hydrogenation".*
  3,540,995; *"H-Coal Process, Slurry Oil System"* (Jointly with U.S. Secretary of the Interior).
  3,607,719;*"Low-Pressure Hydrogenation-Ebullated Bed".*
  3,617,465; *"Expanded Particulate – Solids Bed Hydrogenation".*
  3,617,474; *"Low Sulphur Fuel Oil".*

Institute of Gas Technology.
  3,708,269; *"Fossil Fuel Hydrogasification – Pipeline Gas".*

Kerr–McGee Corporation.
  3,607,718; *"Hydrogenation Solvents".*

Koppers Company Inc.
  2,738,311; *"Hydrogenation (pasting media)".*

Leas Bros. Development Corporation.
  3,779,893; *"Desulphurised Liquids and Gas".*

Lummus Company: The
  3,375,188; *"De-ashing without Hydrogen".*
  3,379,638; *"Ash-Free Anode Carbon"* (Jointly with Great Lakes Carbon Corp.).

Occidental Petroleum Corp.
  3,698,882; *"Continuous Process Conversion – Carbonaceous Solids".*

Phillips Petroleum Co.
  2,756,194; *"Nickel Carbonyl Catalyst".*

Pott-Broche.
  2,308,247; *"Extraction and Hydrogenation of Coal".*
  292,808 (British Patent) 12th July 1927; *"Destructive Hydrogenation".*
  657,409 (French Patent) 27th June 1928; *"Hydrogenating Coal".*

Pyrochem Corporation.
  3,244,615; *"Contact Catalysis".*
  3,247,092; *"Quadri-Phase Hydrogenation".*

Research Council of Alberta, The
  3,505,201; *"Coal-Oil Suspension Separation".*

Secretary of the Interior, U.S.
2,464,271; *"Low-Pressure Hydrogeneration"*
2,476,999; *"Solvent Depolymerization".*
2,606,142; *"Zinc Catalyst Hydrogenation".*
2,860,101; *"Balanced Hydrogenation"*
3,341,447; *"Solvent Processing".*
3,371,049; *"Zinc Halide Regeneration – Polynuclear Hydrocarbons".*
3,355,376; *"Zinc Halide Polynuclear Aromatics (*Con. Coal Company)
3,375,175; *"Coal Pyrolysis"*
3,477,942; *"Fuel Cell Pyrolysis".*
3,516,808; *"Carbon Dioxide Acceptors-Melt Process"* (Con. Coal Company)
3,523,886; *"Solvent Extraction"* (Consolidation Coal Co.)
3,540,995; *"H-Coal Process: Slurry Oil System"* (H.R. Inc.)
3,556,978; *"Hydrogenation with Alumium Chloride".*
3,565,766; *"Coal and Residue Co-Pyrolysis".*
3,567,412; *"Carbonaceous Fuels Gasification".*
3,577,337; *"Quinoline Solubisation and Ultrasonic Treatment".*
3,594,329; *"Zinc Chloride Regeneration-Hydrocracking".*
3,625,861; *"Zinc Chloride Regeneration-Hydrocracking".*
3,629,159; *"Zinc Chloride Regeneration-Hydrocracking".*
3,746,522; *"Gasification of Carbonaceous Solids".*
3,782,913; *"Two-Stage Coal Gasification with Forced Mixing and Steam Treatment of Recycle Char"* *(BI-GAS)*
3,817,723; *"Two-Stage Gasification of Pretreated Coal"* (BI-GAS)

Shell Development Company/Shell Oil Company.
2,392,588; *"Catalytic Treatment".*
3,502,564; *"Metal Sulphide and Napthalene Catalytic Hydroprocessing".*
3,527,691; *"Dual Function Catalytic Absorbent".*
3,532,617; *"Hydroconversion Catalysts".*
3,543,665; *"Freidel-Crafts Catalysts".*
3,549,512; *"Solid Absorbents".*

Squires, A.M.
3,597,327; *"Fluidised Bed Pyrolysing"..*

Standard Oil Development Company.
2,379,077; *"Thermal Cracking".*
2,436,938; *"Motor Fuel".*
2,557,680; *"Fluidised Carbonisation".*
2,676,908; *"Alkali Carbonisation".*
2,697,718; *"Gasoline".*

Stoewener, Keunecke and Becke.
2,337,944; *"Catalysts".*

Sun Oil Company.
3,594,304; *"Thermal Liquefaction–Solvents".*
3,594,305; *"Ebullating Bed Catalytic Hydrogenation".*
3,770,399; *"Coal Gasification Process"* (Sun Research & Development Co.).

Texaco Development Company.
2,572,061; *"Liquid–Phase Hydrogenation".*
2,753,296; *"Liquid–Phase Hydrogenation".*
3,075,912; *"Hydroconversion".*
3,544,291; *"Coal Gasification Process"* (Texaco Inc.).
3,740,204; *"Methane from Carbonaceous Fuels Process".*

Texas Company, The
2,595,979; *"Underground Liquefaction".*
2,658,861; *"Hydrogenation".*
2,664,390; *"Carbonisation".*
2,681,300; *"Hydrogenated Thianapthenes".*

Thomas C.L.
2,377,728; *"Catalyst".*

Total Energy Corp.
3,692,506; *"High BTU Gas Content from Coal".*

Transcontinental Gas Pipe Line Corp.
   3,728,093; *"Synthetic Pipeline Gas".*

Universal Oil Products Company.
   2,406,810; *"Solids Treatment".*
   3,477,941; *"Solvent Extraction".*
   3,488,278; *"Countercurrent Solvent Extraction".*
   3,503,863; *"Liquid Fuels Extraction – Coking".*
   3,503,864; *"Hydrogen Sulphide Solvent Extraction".*
   3,503,865; *"Microwave Liquefaction".*
   3,505,202; *"Solvent Extraction".*
   3,505,203; *"Hydrogen Donor Solvents".*
   3,520,794; *"Shell and Tube Extraction".*
   3,535,224; *"Dual Solvent Extraction".*
   3,536,608; *"Colloidal Size Coal Liquefaction".*
   3,583,900; *"Three-Stage Solvent Extraction".*
   3,598,717; *"Two-Stage Solvent Extraction".*
   3,598,718; *"Two-Stage Solvent Extraction".*

Union Carbide Corporation.
   2,832,724; *"Catalytic Hydrogenation (Coal Paste)".*
   3,084,118; *"Hydrogenation Product Refining".*
   3,171,369; *"Agglomerating Ash Gasification Process".*
   3,231,486; *"Catalytic Hydrogenation – Carbonised Vapors".*

University of Wyoming.
   3,505,204; *"Direct Conversion".*

Winkler, J.
   3,282,826; *"Friable Metal Reactants Depolymerisation".*

   Information presented herein derived from U.S. Patents Digests, O.C.R. Annual Reports and private communications.

APPENDIX III

CLEAN FUEL FROM COAL PROCESS (C.F.F.C.)

Process Developer: The Lummus Company, a subsidiary of Combustion Engineering, Inc., Bloomfield, New Jersey.

Description: Coal is crushed, sized and dried then slurried and solubilized in a process-derived aromatic recycle solvent. The digested coal slurry is hydro-desulphurized by reaction with hydrogen in the presence of a catalyst at temperatures of 750° to 850°F and pressures of 1,000-4,000 p.s.i.g. The hydrotreating catalyst employed may be cobalt molybdate, tungsten-nickel sulphide, nickel molybdate or mixtures thereof. The hydro-desulphurization step occurs in an ebullating-bed reactor. During the catalytic hydro-desulphurization step, nitrogen and oxygen levels in the coal solution are also significantly reduced.

The hydrotreated coal slurry then passes to the de-ashing and fractionation section. In this step, the insoluble material in the liquefied coal slurry is separated from the coal liquefaction product by introducing the coal slurry to a gravity settling zone in the presence of a liquid promoter. This liquid promoter has a 5 volume percent distillation temperature of at least 250°F and a 95 volume percent distillation temperature of at least 350°F and no greater than 750°F. The promoter liquid consists of (at least) one of the following: kerosene, kerosene fractions, middle distillates, gas oils, heavy naphthas and white oils, and possesses a characterization factor of at least 9-75.

The ratio of the liquid promoter to coal liquefaction product is selected so as to give from the gravity settling zone an essentially insoluble-free overflow and an underflow which contains the insoluble material. The liquid promoter is then separated from liquid overflow in a recovery zone and recycled to the gravity settling zone. The underflow, containing insoluble residue, is withdrawn from the separation zone and sent to a stripping unit where a fraction boiling below 900°F is recovered and passed to the product recovery zone. A recycle liquefaction solvent stream is separated from the product recovery unit and sent to the coal slurrying zone.

The primary product from the product recovery zone, after being stripped of recycle solvent, is a low-sulphur liquid similar to petroleum derived, low-sulphur fuel oil. The mineral-rich residue from the stripper bottoms may be calcined, coked, or gasified to provide hydrogen for the process.

Operating Conditions:

Process Section	Temp. °F	Press, p.s.i.	Reactants	Product
Solvation	800-850	1,000	Coal-solvent	Coal sol.
Hydro-Desulphurization	750-850	1,000-4,000	Coal slurry-H_2	Slurry

Status: C-E Lummus has developed this process over several years, concentrating specifically on the production of a synthetic liquid fuel oil for power generation use. The process is able to offer the industry a substitute fuel oil that meets the air quality sulphurous and particulate emussion standards without resorting to costly stack-gas cleaners.

The major features claimed for the C.F.F.C. process lies with the ability to produce a synthetic fuel oil from coal with minimum hydrogen consumption by the utilization of catalytic hydro-liquefaction during which process the nitrogen, as well as the sulphur and oxygen levels are significantly reduced without undue severity in the liquefaction reactor.

Another important feature of the C.F.F.C. process is the de-ashing step employed to separate insoluble residue (ash and unreacted coal) from the liquid product slurry.

C-E Lummus has recently concluded tests in their Bloomfield, N.J. pilot plant, which successfully demonstrated the removal of insoluble residue from un-de-ashed product from the Wilsonville S.R.C. pilot plant. These tests were carried out by Lummus for the Electric Power Research Institute, a co-sponsor of the Catalytic Inc. S.R.C. plant. (See S.R.C. – PAMCO.)

To date, combustion tests on C.F.F.C. synthetic fuel oil have indicated that the product exhibits combustion characteristics of a similar nature to those of residual petroleum oils.

Other potential uses for the C.F.F.C. product are for the production of aromatic chemicals, carbon products and derivatives, and high-quality cokes.

Data generated by the operation of the C.F.F.C. pilot plant have been employed as the basis for a conceptual design of a commercial-scale coal liquefaction plant. Estimates of plant input requirements for a plant producing 25,000 barrels per stream day of 0.5 weight percent sulphur, 0.1 weight percent ash, clean fuel oil from Illinois No. 6 coal are:

Raw coal	11,400 short tons/day
Water	3,600 gal/min
Power	22,400 kw

In December, 1974, three U.S. patents were issued to C-E Lummus:

3,852,182 of December 3, 1974
3,852,183 of December 3, 1974
3,856,675 of December 24, 1974

Applications for patents have also been lodged in numerous foreign countries.

BURLINGTON AND NORTHERN RAILROAD CO.

The Lurgi and Koppers-Totzek processes are under evaluation by B. & N. Railroad Co. for the production of synthesis gas for ammonia in their proposed coal-based fertilizer plant at Circle, Montana. The company has filed a 10-year plan with the Montana Natural Resources Dept. detailing estimates of requirements of 9,000 ton per day of lignite and up to 9,000 acre feet per year of water for the coal-ammonia plant. Target date for the commencement of construction of the ammonia plant is the third quarter of 1976.

Burlington and Northern's other plans include construction of a methanol-from-coal plant to be underway in 1979-80 and a synthetic diesel oil-from-coal plant to be constructed in the period 1983-85. All these coal conversion plants will be located near Circle, Montana.

CARBON BLACK (Institute of Gas Technology)

The Chicago-based Institute of Gas Technology has developed a process for the catalytic production of carbon black (<1% ash and <0.25% S) from coal. The I.G.T. process operates on air alone as the oxidant, and utilises a CO-rich gas and the water gas-shift reaction.

Initially, the coal is devolatilised to generate a stream of pyrolysis tar/gas products. The devolatilised char is then gasified to produce a gas stream of CO and H_2 which is cooled by passage through a fluidised bed of catalysts which cause precipitation of finely-divided carbon to which may be added an admixture of the pyrolysis-product stream.

The process is carried out at pressures of 880–1,100 p.s.i.

COALCON HYDROCARBONIZATION

The following companies have joined the industry group being formed by the Coalcon Co. to develop the Coalcon Hydrocarbonization 2,600 T.P.D. demonstration plant:—

Atlantic Richfield Co.
Mobil Research and Development Corp.
Martin-Marietta Corp.

This brings the industry team to nine.

In addition to these companies, there are five state and county agencies in Ohio and West Virginia involved. There are six states vying for the demonstration plant to be located within their regions:

Illinois; Indiana; Kentucky; Ohio; Pennsylvania and West Virginia.

Wood County, West Virginia and Belmont County, Ohio appear to be the front-runners, however, the other contenders are submitting competitive proposals. Site selection for the Coalcon plant is to be made by 1st January, 1976.

COAL LIQUEFACTION PROCESSES—COST MODELLING
(West Virginia University)

The West Virginia University, Morgantown, W.Va., has completed an economic comparison of three major coal liquefaction processes using a computerized model of each of the processes evaluated, i.e., S.R.C., H-COAL and SYNTHOIL.

Independent of the parameter investigated (oil cost, capital investment, operating cost, power cost etc.) the S.R.C. process costs were consistently lower than those for Synthoil or H-Coal.

The work was sponsored by the National Science Foundation (now under E.R.D.A.).

COALPLEX (N.C.B.)

The British National Coal Board has proposed a development program for the period 1975-79 which would research the coalplex concept of integrating coal liquefaction, gasification and pyrolysis plants with power generation plants. The overall cost of the program was estimated at £19,000,000 which is to be funded by the European Coal and Steel Community.

C.O.E.D. (F.M.C. Corporation)

The F.M.C. Corporation has made plans for shipment of 1,000 tons of C.O.E.D. char to Spain where its performance will be tested in a Koppers-Totzek gasifier at a commercial fertilizer plant. The tests will be sponsored by E.P.R.I. and E.R.D.A. and the results will be evaluated by the Koppers Company. (See p. 7 of this book.)

EL PASO NATURAL GAS CO.

As a result of the increasing work-load on El Paso's Burnham Coal Gasification Complex in the Four Corners area of New Mexico, (expected to come on-stream in mid-1978), the company has been forced to indefinitely postpone its proposed coal gasification complex in Dunn, Bowman and Stark Counties, North Dakota. El Paso has withdrawn its application for State water rights in relation to the four proposed North Dakota S.N.G. plants but is continuing to assemble coal reserves in the three counties for future utilisation in S.N.G. plants.

E.R.D.A.

The recently-created Energy Research and Development Administration (E.R.D.A.) incorporates the former energy R.&D. functions of the organisations listed on the following page.

Atomic Energy Commission (excluding the nuclear reliability and safety monitoring functions which are assumed by the new Nuclear Regulatory Commission);

Office of Coal Research, U.S. Dept. of the Interior;

Bureau of Mines, U.S. Dept. of the Interior;

National Science Foundation; and

Environmental Protection Agency.

EXXON GASIFICATION

EXXON Research and Engineering Co. is developing a new approach to S.N.G. production with a catalytic coal hydrogasification process.

In this process, pulverized coal is reacted in a fluidized bed with steam and hydrogen in the presence of an alkali-metal catalyst. Methane produced in the hydrogasification reaction is cryogenically separated from the offgases and can be used directly as a pipeline gas. Carbon monoxide, cryogenically separated, is steam-reformed to hydrogen which is recycled, with the hydrogen stream from the cryogenic separation unit, to the hydrogasification reaction zone. The hydrogen and steam concentrations are controlled to balance the exothermic and endothermic reactions, thus providing a thermoneutral process for hydrogasification which optimises methane production.

GEGAS

General Electric and the Electric Power Research Institute have announced plans to construct a 1 T.P.H. pilot plant to evaluate the GEGAS gasification process for the production of low-Btu gas from a wide range of impure caking coals. This pilot plant, to produce 100,000 SCF per hour of 150 Btu gas, will be located at General Electric's Research and Development Center in Schenectady, N.Y.

H-COAL (Hydrocarbon Research Inc.)

The Hydrocarbon Research Inc. Development Laboratory at Trenton, N.J., has reported significant success in the removal of ash and inorganic sulphur from the H-COAL liquid by the use of magnetic separation. To date, filtration, solvent precipitation and magnetic separation have been tested in the solid-liquid separation step of the H-COAL process at Trenton. The economics of magnetic separation will be evaluated in 1975.

INDIA

The Indian Central Fuel Research Institute conducts studies related to:—

1. transportation of coal in a concentrated form to reduce delivery costs
2. coal conversion to liquid fuels, S.N.G., low-Btu gas, chemical feedstocks and domestic fuel
3. processing of coal for utilisation in the iron and steel industry

Specific research related to coal conversion includes:—

a) Low-temperature carbonisation

Two continuous vertical retorts are operated of 20-30 tonnes per day total capacity. Operation of these retorts is providing information for the construction of a 1,500 T.P.D. carbonisation plant in West Bengal by C.F.R.I., in conjunction with Engineering Projects India Ltd. This plant will produce 13-15 million S.C.F./day of gas and 1,000 T.P.D. of coke.

b) Formed-coke

Operation of a 1 tonne/hour pilot plant for the production of formed-coke from noncoking coals by fluidised-bed carbonisation has culminated in the construction of a 300 T.P.D. formed-coke plant in Talchar (Orissa State).

c) Gasification

R.&D. on steam-oxygen gasification entails the operation of:—a Lurgi plant of 0.8 tonne/day capacity, a

Koppers-Totzek gasifier of 110 kg/hour capacity, and a slagging gasifier. The emphasis at present is on the application of solid heat carriers due to the high ash content of Indian coals.

d) Utilisation of Coal-Distillation Products

A variety of processes have been developed at C.F.R.I. for the extraction of valuable chemical products from coal tars.

e) Direct Conversion

An 18 kg/hour pilot plant for the hydrogenation of coal will be operational in the near future with plans in hand for a 24 T.P.D. demonstration plant to follow. This work is co-sponsored by the National Research Development Corp. of India and the Indian Oil Corp.

f) Industrial Carbon

A 1 T.P.D. plant to produce activated carbon from lignite has been constructed in Madras State. The manufacture of electrode carbon from coal tar is being actively researched.

g) Catalysts

Research into hydrogenation, hydro-refining and Fischer-Tropsch synthesis catalysts is continuing.

h) Acetylene Production

Studies are in progress on the production of acetylene from coal by plasma and laser-irradiation pyrolysis.

i) Energy Generation

Emphasis is being placed on various aspects of energy generation and storage including:—

1. fuel cells based on methanol and oxygen
2. MHD generation based on Lurgi, Koppers-Totzek, producer-gas and water-gas generators.

"IN SITU" GASIFICATION AND LIQUEFACTION (Belgium)

Due to the existence of very large reserves (of the order of 10,000 million tons) of coal in the Campine area of the Belgian Southern Coal Basin which are not amenable to economic extraction, the Belgian National Institute of Extractive Industries has now turned its attention towards the application of the "in situ" gasification, liquefaction and degasification techniques to utilise these resources. The "in situ" gasification process is currently being tested in the deeper seams of the region combined with the extraction of the methane which ascends into the upper portions of the deposits. This technique employs the introduction of gasifying media at pressures up to 1,500 p.s.i. via drill holes spaced 600 feet apart which penetrate seams at a depth of 2,000 feet. The strata between the coal seams is heated to a high temperature and acts as a heat sink where the heat is utilised for producing synthesis gas during the gasification process and for the generation of steam following the gasification process. Four other drill holes are to be put down for the recovery of the methane from the upper part of the deposit.

Due to the interest in "in situ" technology, the Belgian Institute will be presenting data on its field tests to the European Economic Community Commission.

LURGI

In the Federal German Republic, a large-scale project to further develop Lurgi pressure gasification and the production of synthetic liquid and gaseous fuels from coal has been undertaken by a consortium of oil, gas and electricity companies.

METHANATION (Conoco Coal Development Corp.)

Following the successful steady-state operation of the jointly-designed Conoco and British Gas Corp. fixed-bed, gas-recycle, methanation reactor at Westfield in Scotland, bench-scale reactors continue to be operated both in Scotland and at Conoco Coal Development Co.'s Ponca City, Okla., laboratory to augment the data on long-term catalyst durability.

METHANATION (Ventron Corp.)

The Ventron Corp. has developed a process for the production of high-purity methane from synthesis gas over a transition-metal catalyst (nickel, palladium or platinum) intercalated in a graphite structure in the metallic state. The significant feature of this methanation process is that the reaction can be carried out at pressures as low as 1 atmosphere and within a temperature range of 400°-850°F. The catalyst is sulphur-tolerant up to 0.3% S.

METHANOL-COAL REACTION PROCESS (Texas Tech. University)

Researchers at Texas Tech. University, Lubbock, Tex., have conducted experiments in a 1 litre stirred batch autoclave to demonstrate the feasibility of employing methanol as a coal-conversion medium. The process involves the slurrying of coal in methanol with subsequent hydrogenation of the coal through the thermal decomposition of the methanol.

The experimental results indicated that methanol is uniquely absorbed by coal which provides an intimate dissemination of the hydrogen source within the coal particles. The thermal decomposition of the methanol within the coal particles yields highly-reactive free radicals for the hydrogenation reactions. The final reaction product of the decomposition of methanol is carbon monoxide and this could, in turn, react with the moisture in the coal to generate additional nascent hydrogen by the water-gas shift reaction.

Coal conversions of up to 41% (d.a.f.) at 600°C and 1,400 p.s.i. were obtained, with a significant increase in the rate of coal conversion compared with gaseous-phase hydrogenation due to the elimination of boundary-layer resistance.

METHANOL—M.T.B.E. SYNTHESIS

The Sun Oil Co. has evaluated many oxygenated hydrocarbons (alcohols and ethers) for utilisation as high-octane motor fuels and has selected methyl tertiary butyl ether (M.T.B.E.) for further evaluation as an additive to unleaded gasolines in concentrations up to 15 volume percent. Sun Oil has presented a concept for a fully-integrated plant to produce a mixture of M.T.B.E. and alcohols from coal via catalytic synthesis of coal-derived synthesis gas. The route incorporates the catalytic synthesis of a controlled mixture of alcohols (methanol, isobutanol and higher alcohols) with subsequent dehydrogenation of the isobutanol to isobutene which is reacted with the methanol to produce M.T.B.E. Approximately 2.0 barrels of M.T.B.E. per ton of coal can be produced by the process with 0.2 barrels per ton of hydrocarbons and 0.7 barrels per ton of higher alcohols as by-products.

METHANOL SYNTHESIS—LURGI

The first application of Lurgi's low-pressure methanol technology in the United States will be a 1,000 ton/day plant scheduled for completion in 1976 at Celanese Chemical's Bishop, Texas, complex. The synthesis-gas feed may be prepared by reforming of natural gas or naphtha, or coal gasification.

NEW SOUTH WALES

The New South Wales State Government has appointed Gas Developments Corporation, an affiliate of the Chicago Institute of Gas Technology, as consultants in the field of coal conversion.

NUCLEAR-WASTE-HEAT GASIFICATION

In West Germany, Union Kraftstaff and Rheinische Braunkohlenwerke A.G. (the developers of the Rummel Slag Bath Single Shaft process), are constructing a pilot plant at Wesseling to evaluate the use of nuclear reactor process heat in the gasification of coal in fluidized beds to produce synthesis gas and methane.

Others involved in researching the application of nuclear process heat to endothermic coal gasification are:—

      Stone & Webster Engineering Corp. and Gulf General Atomics Co.
        (see "Solution-Gasification")
      Bergbau-Forschung GmbH of Essen

Kernforschungsanlage Julich GmbH of Julich
Coal Mining Research Centre, Essen.

The R.&D. in this field is sponsored in the German Federal Republic by Bundeministerium fur Forschung und Technologie.

The Coal Mining Research Centre of Essen considers that a 3,000 MW nuclear-waste-heat coal gasification plant could be operational by 1990. Such a plant would process 12,000 tonnes per day of coal to produce 300 million cubic feet per day of S.N.G. and 500 MW of power.

These figures are interesting in that the U.S. Lurgi based S.N.G. plant estimates indicate a yield of 250 million S.C.F./day of S.N.G. from a coal feed rate of 25,000 T.P.D. or, equivalently, 300 million S.C.F./day for 30,000 T.P.D. of coal. The difference of 18,000 T.P.D. presumably represents the amount of coal required to supply the heat for the endothermic coal-steam gasification reaction in the Lurgi process which would be supplied by the nuclear process heat in the C.M.R.C. process. Since the utilisation of the nuclear-waste-heat does not detract from the operational efficiency of the H.T.R., the considerable benefits of integrating coal gasification plants with H.T.R. power plants is obvious, especially hydrogasification units.

The Centre has been working on this technique for a number of years and has developed plans for an experimental high-temperature, gas (He)—cooled nuclear reactor of 750 MW coupled with a 1,200 T.P.D. gasification plant which will produce an overall power equivalent of 900 MW. Such a plant would be expected to be operational by 1983 at a cost of A$1,100 million however this would be dependent upon the successful operation of a 300 MW prototype high-temperature, gas-cooled nuclear reactor now under construction at Ventrop-Schmelhausen.

OAK RIDGE HYDROCARBONIZATION PROCESS

The Oak Ridge National Laboratory has been awarded a U.S. $600,000 contract by E.R.D.A. to provide experimental engineering support data for the Hydrocarbonization process. Phase I of the program is to consist of:—

1. evaluation of available data on the process
2. the operation of experimental batch units to test critical process steps and to provide design data for a continuous bench-scale process unit
3. design and construction of the continuous bench-scale facility of about 10 lb/hour capacity.

Phase I is to be completed by the end of 1975.

In Phase II, the continuous bench-scale unit will be operated to study the effects of process variables under a wide range of operating conditions. The goal is to fully define the potential of:—

a) noncatalytic hydrocarbonization
b) process steps employing supported catalysts with molten-salt components to produce a range of desirable products.

PANHANDLE EASTERN PIPE LINE CO.

Panhandle Eastern Pipe Line Co. will commence construction of their first 250 MM S.C.F./day S.N.G. plant based on Lurgi coal gasification plus methanation, in late 1976 or early 1977. The plant will be located about 15 miles north-east of Douglas, Wyoming.

In another related development, Panhandle Eastern has been allocated 26,500 acre feet per annum of water from the North Platte River for its proposed 250 million S.C.F./day S.N.G. plant.

POLAND

The Main Institute of Mining in Poland co-ordinates the research and development of coal conversion processes in concert with about 20 research establishments in Czechoslovakia, The German Democratic Republic, Rumania, Hungary and the Soviet Union. Recently, this technical co-operation has been extended to include agreements for exchanges on the gasification and liquefaction of coal with the United States and the Federal Republic of Germany.

SASOL I

To handle any short-term energy shortages in South Africa until SASOL II comes on-stream, SASOL is immediately initiating a U.S. $65,000,000 expansion program at SASOL I to increase the capacity of the Lurgi gasification battery by 40 per cent to double the supply of fuel gas to industry. The new gasifiers will be of a size unequalled in the world.

S.R.C.—SASOL

The research laboratories of The South African Coal, Oil and Gas Corp. Ltd. (SASOL), have been conducting extensive research into the Solvent Refined Coal (S.R.C.) Process. A pilot plant, in operation since 1971 has provided sufficient data to establish optimum operating conditions in critical process steps such as the degree of solvent extraction, filterability, and the characteristics and properties of the S.R.C. product.

S.R.C.—MITSUI/GULF

The Gulf Oil Corp. and the Mitsui Group announced on 19th June 1975 that an agreement was entered into (December 1974) which aims for co-operation in commercializing Gulf's Solvent-Refined Coal (S.R.C.) process. The Gulf Oil subsidiary, Gulf Minerals and Energy Co. (GEMCO) will supply patent licences and technical support and Mitsui will undertake the bulk of the financing. Under the agreement, the partners will exchange technical data, jointly improve the economics and efficiency of the S.R.C. process and construct and operate commercial S.R.C. plants. (Refer to S.R.C.—MITSUI page 29 and S.R.C.—PAMCO page 30.)

SUPER-CRITICAL AQUEOUS EXTRACTION (Bechtel)

Bechtel International Corporation is developing a process for the liquefaction of coal in an aqueous medium. Pulverised coal is slurried in an aqueous medium and contacted with super-critical water in the presence of hydrogen to depolymerise the coal rapidly to aralkanes and gaseous hydrocarbons.

VICTORIA

The Victorian State Cabinet approved on 19th May, 1975, the establishment of a new Committee within the Ministry of Fuel and Power to advise the Government on coal conversion activities. The Committee will be comprised of representatives of the Gas and Fuel Corporation and the State Electricity Commission, as well as experts from industry and the academic world.

WELLMAN GALUSHA

A proposal has recently been announced for M.W. Kellogg Co. and the College of Earth and Mineral Sciences of the Pennsylvania State University to jointly develop a demonstration plant to produce medium-Btu gas suitable for methanation to pipeline quality based on a modified Wellman-Galusha unit.

APPENDIX IV

COAL CONVERSION PROCESSES
for the production of
Liquid, Solid and Gaseous Products

NOTE: The processes represented in these charts refer to significant techniques undergoing current research and development evaluation. A more complete list of suitable processes can be found in the body of the book.

CLEAN-COKE

PROCESS UNIT	TEMP. (°F)	PRESSURE (psi)	PRODUCTS
CARBONIZATION	1200–1400	100	LIQUIDS, GAS, CHAR FOR COKER
HYDROGENATION	900	3000–4000	LIQUIDS, GAS, FILTER CAKE FOR GASIFIER

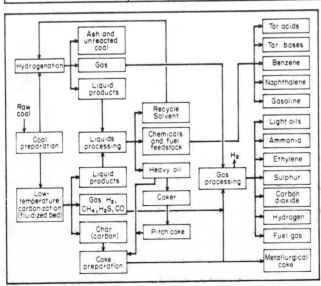

Testing in 500 lb. per day integrated process development units at Monroeville, Pa., is continuing with sufficient experimental work completed for the preliminary design of a 100 ton per day pilot plant. Estimates have been presented on the yield of products from a commercial plant processing 17,000 ton per day.

COALCON HYDROCARBONIZATION

NO DETAILS ARE AVAILABLE ON OPERATING CONDITIONS

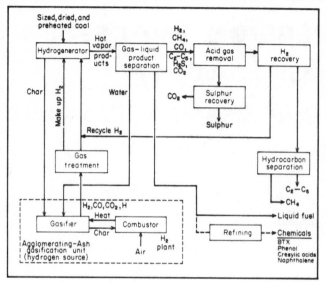

The process has been demonstrated in a 25 ton per day pilot plant and a 300 ton per day semi-works plant. Design of a 2600 ton per day demonstration plant is underway with construction scheduled for completion in 1979. This demonstration plant is being designed to produce 3900 barrels per day of liquid products and 22 million S.C.F. per day of pipeline gas.

C.O.E.D.

PYROLYSIS SECTION	TEMP. (°F)	PRESSURE (psi)	PRODUCT
FLUIDIZED BED No. 1	600	5-10	
" " No. 2	850	5-10	
" " No. 3	1000	5-10	CHAR, OIL, GAS, LIQUOR
No. 4	1500	5-10	
HYDROTREATMENT SECTION	TEMP. (°F)	PRESSURE (psi)	PRODUCT
CATALYTIC FIXED-BED	750	2500-3100	SYNTHETIC CRUDE OIL

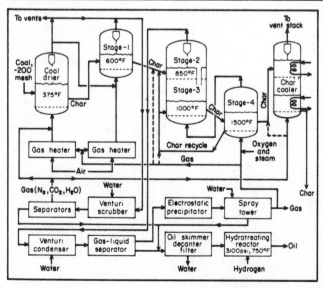

A 36 ton per day pilot plant at Princeton, N.J. has been successfully demonstrating the process since 1970 in long-duration runs of over 30 days. The remaining development work on the C.O.E.D. project consists of performing large-scale char gasification tests in commercially-available gasifiers.

C.O.G.

LIQUEFACTION (S.R.C.) SECTION	TEMP. (°F)	PRESSURE (psi)	PRODUCT
SOLVATION-HYDROGENATION	850	1000	LIGHT OIL, S.R.C.
GASIFICATION SECTION	TEMP. (°F)	PRESSURE (psi)	PRODUCT
TWO-STAGE ENTRAINED FLOW	3000	200	SYNTHESIS GAS

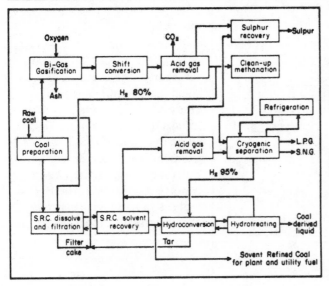

An economic evaluation and conceptual design of the C.O.G. refinery concept have been completed. Individual process sections are currently being tested at Pamco's 50 ton per day S.R.C. pilot plant at Ft. Lewis, Wash. and the BI-GAS project's 120 ton per day pilot plant at Homer City, Pa.

COGAS

PYROLYSIS SECTION	TEMP. (°F)	PRESSURE (psi)	PRODUCTS
FLUIDIZED BEDS (4)	600-1500	5-10	CHAR, OIL, LIQUOR, GAS
HYDROTREATMENT SECTION	TEMP. (°F)	PRESSURE (psi)	PRODUCT
CATALYTIC FIXED-BED	750	2500-3100	SYNTHETIC CRUDE OIL
GASIFICATION SECTION	TEMP. (°F)	PRESSURE (psi)	PRODUCT
GASIFIER-COMBUSTOR	1500-1700	0-30	SYNTHESIS GAS

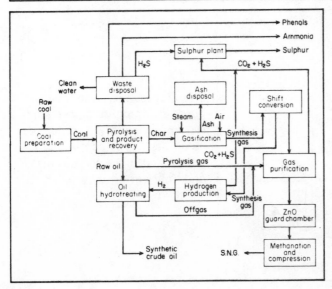

The pyrolysis section of the process is fashioned after the C.O.-E.D. process, which is undergoing pilot plant trials in a 36 ton per day unit at Princeton, N.J. Two char gasification processes are currently being investigated:
1. The Inert Heat-Carrier process in a 2·5 ton per day pilot plant at Princeton, N.J.
2. The Char Heat-Carrier process in a 50 ton per day pilot plant at Leatherhead, England. Both char gasification processes have operated successfully to date.

CONSOL SYNTHETIC FUEL

REACTOR	TEMP. (°F)	PRESSURE (psi)	PRODUCTS
EXTRACTION	765	150	VAPOURS, LIQUID, RESIDUE SLURRY
CARBONIZATION	925	10	CHAR, PYROLYSIS LIQUID & GAS
HYDROTREATMENT	800	3000	NAPHTHA, OIL, GAS, RESIDUE

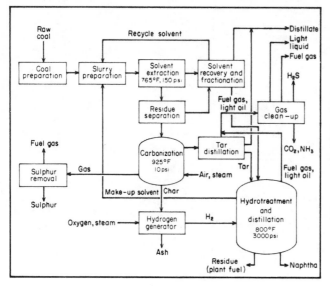

A 20 ton per day pilot plant was operational at Cresap, W. Va., for 35 months to April, 1970, when it was shut down for a study of the process and operating problems. Late in 1974, O.C.R. awarded a U.S.$13M contract to Fluor Corp. to undertake the revamping of the plant over 3 years. Elements of the H-Coal process may be incorporated in the revamped plant.

COSTEAM

REACTOR	TEMP. (°F)	PRESSURE (psi)	REACTANTS	PRODUCTS
STIRRED EXTRACTION	800	4000	COAL-CO-STEAM	FUEL OIL, GAS

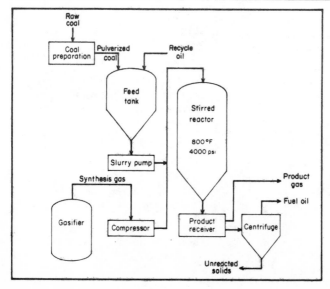

The process has been investigated in batch-autoclave studies employing a 2·2 gallon reactor. Testing is continuing to provide data for the design and fabrication of a pilot plant.

EXXON LIQUEFACTION

REACTOR	TEMP. (°F)	PRESSURE (psi)	REACTANTS	PRODUCTS
SOLVENT EXTRACTION	750	350	COAL – H-DONOR SOLVENT	LIQUID PRODUCTS, GAS
CATALYTIC HYDROGENATION	800	2000	COAL – H-DEPLETED SOLVENT – H_2	LIQUID PRODUCTS, REGENERATED SOLVENT, GAS

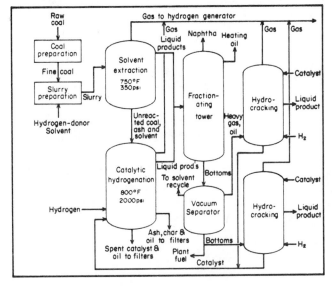

The process stages have been investigated in several continuous pilot plants up to 0·5 ton per day in size since 1967. General process feasibility has been established. Design of a 200 ton per day pilot plant to demonstrate critical process features for scale-up to a commercial plant is underway. This pilot plant will be located near Exxon's refinery at Baytown, Texas.

FISCHER-TROPSCH SYNTHESIS

SYNTHESIS TYPE	CATALYST	TEMP. (°F)	PRESSURE (psi)	PRODUCTS
ARGE FIXED-BED	IRON / COBALT	450	360	PETROL, L.P.G., ALCOHOL, OILS, WAX, GAS, CHEMICALS
KELLOGG FLUIDIZED-BED (SYNTHOL PROCESS)	IRON	620	330	PETROL, ALCOHOL, OILS, WAX, GAS

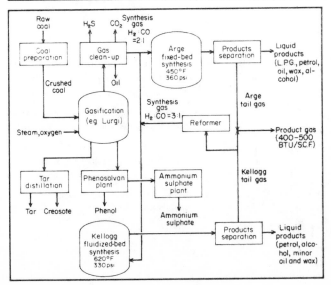

The Fischer-Tropsch synthesis route for the production of a wide range of liquid, solid and gaseous products from coal-derived synthesis gas has been in large-scale commercial operation at Sasolburg, South Africa since 1954. The process was utilized in Germany in World War II for the production of synthetic fuels.

GARRETT'S COAL PYROLYSIS

REACTOR	TEMP. (°F)	PRESSURE	REACTANTS	PRODUCTS
PYROLYZER	1100	ATMOSPHERIC	COAL – HOT CHAR	TAR, PRODUCT CHAR, GAS (700 BTU/S.C.F.)
CHAR HEATER	1200–1600	ATMOSPHERIC	CHAR – AIR	HOT CHAR FOR PYROLYZER

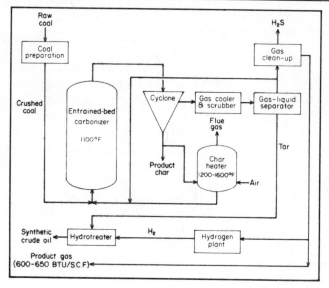

The process was demonstrated in a small, semi-continuous reactor with results indicating a coal-liquid conversion twice that predicted by Fischer Assay. At La Verne, Calif., a 3·6 ton per day pilot plant has been in operation for some years.

H-COAL

REACTOR	TEMP. (°F)	PRESSURE (psi)	REACTANTS	PRODUCTS
CATALYTIC, EBULLATED-BED	850	2250–2700	COAL – H_2 – OIL	SYNTHETIC CRUDE OIL & GAS

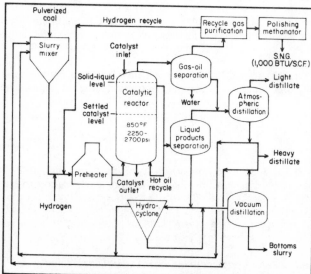

The process has undergone over 4,000 hours of testing in a 3 ton per day P.D.U. and over 40,000 hours of 25–100 lb. per day bench-scale testing at Trenton, N.J. In August, 1974, O.C.R. awarded a contract to H.R.I. for construction and operation of a 600 ton per day pilot plant which will be located at Catlettsburg, Ky.

LURGI-RUHRGAS

REACTOR	TEMP. (°F)	PRESSURE	REACTANTS	PRODUCTS
FLASH-CARBONIZER	1100	ATMOSPHERIC	COAL-HOT CHAR	CHAR, TAR, GAS (700-850 BTU/S.C.F.)

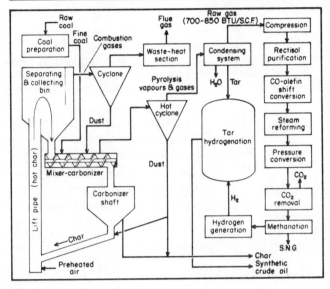

The process was initially tested in a continuous laboratory unit with results indicating tar yields between 129 and 212 per cent of those predicted by Fischer Assay. These results were confirmed in a 12 ton per day pilot plant. Several commercial plants are currently in operation.

METHANOL SYNTHESIS

REACTOR	CATALYST	TEMP. (°F)	PRESSURE (psi)	REACTANTS	PRODUCT
CATALYTIC	CHROMIUM, ZINC OR COPPER OXIDE	500	750-4500 (Depending on catalyst)	$2H_2 : 1 CO$	METHANOL

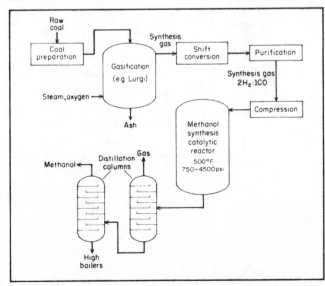

Methanol Synthesis via coal gasification has been utilized commercially for many years, primarily based on Lurgi and Koppers—Totzek gasification. The suitability of the Synthane process for methanol synthesis is under study by the U.S. Bureau of Mines. Any process which yields an H_2/CO synthesis gas with an H_2 to CO ratio of 2:1 after shift conversion can be utilized for methanol synthesis.

PROJECT LIGNITE (S.R.L.)

REACTOR	TEMP. (°F)	PRESSURE (psi)	PRODUCT
EXTRACTION-HYDROGENATION	750–840	1500	PRODUCT SLURRY, GAS
NO DETAILS ARE AVAILABLE ON OPERATING CONDITIONS FOR THE OTHER PROCESS STEPS			

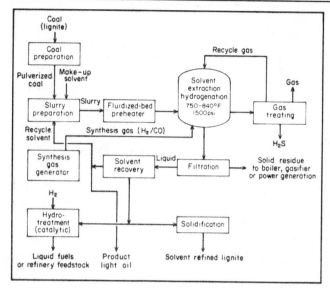

The first stage of a 50 lb. per hr. P.D.U. is in operation at The University of North Dakota, Grand Forks, N.D. This stage produces a solid, solvent refined lignite. A second stage, to be operational in early 1976, will upgrade the S.R.L. to liquids by fixed-bed catalytic hydrogenation. The P.D.U. results will provide design criteria for a larger pilot plant.

SOLVENT EXTRACTION-U.O.P.

REACTOR	TEMP. (°F)	PRESSURE (psi)	REACTANTS	PRODUCTS
SOLVENT EXTRACTION	700–1000	500	COAL-SOLVENT-H_2	SYNTHETIC CRUDE OIL & GAS

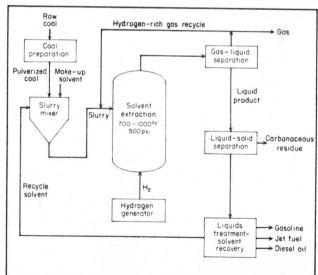

The process has been tested for two years on P.D.U. and pilot plant scales at Des Plaines, Ill., and plans are under consideration for the construction of a demonstration plant.

S.R.C.- C.F.C. PROCESS

NO DETAILS ARE AVAILABLE ON OPERATING CONDITIONS

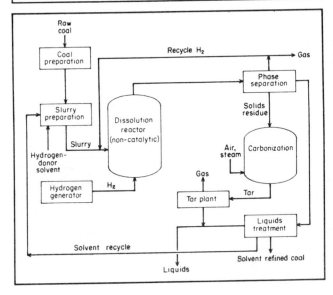

A proposal for a 900 ton per day demonstration plant at Toledo, Ohio, is under consideration. The plant is to combine elements of the Ft. Lewis and Wilsonville S.R.C. pilot plants together with elements of the Cresap C.S.F. pilot plant to produce 500 tons per day of S.R.C. and 400 barrels per day of low-sulphur distillates. The CO_2 Acceptor process may also be utilized in the plant's gasification route.

S.R.C.-PAMCO

REACTOR	TEMP. (°F)	PRESSURE (psi)	REACTANTS	PRODUCT
DISSOLVER	815	1000	COAL–SOLVENT–H_2	S.R.C.

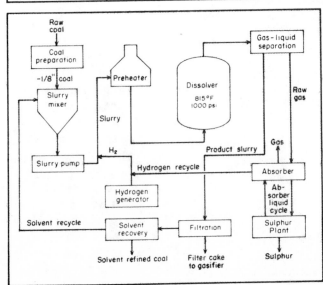

A 6 ton per day pilot plant has been in operation at Wilsonville, Ala., for some years to provide design data for a 50 ton per day pilot plant at Ft. Lewis, Wash., which started up in October, 1974. The Ralph M. Parsons Co. is designing a commercial-scale plant under contract to O.C.R. Wheelabrator-Frye, Inc. and Southern Services, Inc. plan to build a 1,000 ton per day demonstration plant to utilize the S.R.C. process. This plant will be designed for expansion to produce 10,000 tons per day of S.R.C.

SYNTHOIL

REACTOR	TEMP. (°F)	PRESSURE (psi)	REACTANTS	PRODUCTS
CATALYTIC-FIXED-BED	850	2000 – 4000	COAL–H_2–OIL	FUEL OIL, GAS

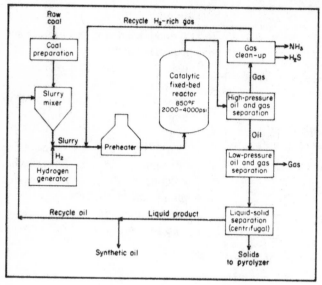

A U.S.$6·9M contract has been let to Foster-Wheeler Corp. for the design and engineering of a 10 ton per day pilot plant to be located at Pittsburgh, Pa. The plant is being designed to yield 24 barrels per day of low-sulphur oil.

TOSCOAL

REACTOR	TEMP. (°F)	PRESSURE	REACTANTS	PRODUCTS
PYROLYZER	800–1000	ATMOSPHERIC	COAL–HEAT	CHAR, OIL, GAS

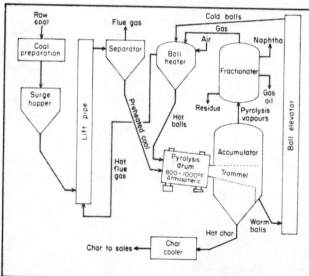

A 25 ton per day pilot plant at Rocky Flats, Colo., has been in operation for some years. Pilot plant results indicated tar yields comparable to those predicted by Fischer Assay. The process is similar to the Tosco II oil shale retorting process and both can be represented by the same flow sheet. Testing of the Tosco II process in a 1,000 ton per day semi-works plant is complete. Construction is underway on a 66,000 ton per day (oil shale) commercial plant at Parachute Creek, Colorado.

114

COAL GASIFICATION PROCESSES
for the production of
Substitute Natural Gas

Those processes requiring an "outside" methanation step to upgrade the raw gas to pipeline quality could utilize a commercially-proven methanation process, e.g., "Hot-Gas Recycle."

The following high-Btu processes could also be utilized for the production of low-Btu fuel gas by the substitution of air for oxygen: BI-GAS, GEGAS, HYGAS (oxygen), LURGI, MOLTEN SALT, PATGAS, SYN-THANE, WELLMAN-GALUSHA and WINKLER.

NOTE: The processes represented in these charts refer to significant techniques undergoing current research and development evaluation. A more complete list of suitable processes can be found in the body of the book.

AGGLOMERATING ASH

FLUIDIZED BED	TEMP. (°F)	PRESSURE (psi)	REACTANTS	PRODUCT GAS
COMBUSTOR	2000 – 2100	100	CHAR – AIR	LOW BTU GAS
GASIFIER	1800	100	COAL – STEAM	MED. BTU GAS

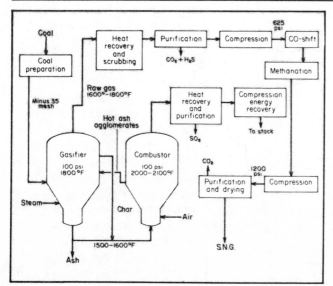

At West Jefferson, Ohio a 25 ton per day pilot plant is due for start-up in early 1975. The pilot plant is to produce 800,000 S.C.F. per day of H_2/CO synthesis gas. No shift or methanation steps will be involved, but these can be added to upgrade med. Btu offgas to S.N.G. at a later stage.

ATGAS - PATGAS

PROCESS	TEMP. (°F)	PRESSURE (psi)	REACTANTS	PRODUCT GAS
PATGAS	2500	50	COAL-O_2-STEAM -LIMESTONE	315 BTU/S.C.F (RAW GAS)
ATGAS	2500	50	AS ABOVE PLUS METHANATION	940 BTU/S.C.F

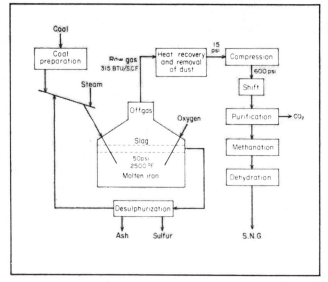

The process has been demonstrated in short-duration runs in a 24 inch internal-diameter gasifier. Future plans call for further development work involving a larger gasifier.

BI-GAS

GASIFIER	TYPE	TEMP. (°F)	PRESSURE (psi)	REACTANTS	PRODUCT GAS
STAGE 2(UPPER)	ENTRAINED FLOW	1700	1000—1500	COAL — STEAM	378 BTU/S.C.F. (RAW GAS)
STAGE I (LOWER)	VORTEX FLOW	2700	1000—1500	CHAR-O_2-STEAM	

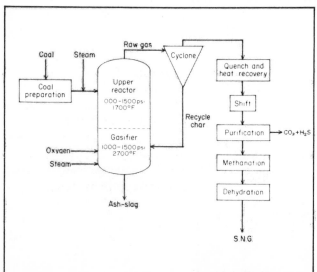

A 120 ton per day pilot plant is under construction at Homer City, Pa. Due for completion in mid-1975, the pilot plant will incorporate a fluidized-bed methanator to up-grade the 378 Btu/S.C.F raw gas to pipeline quality.

CO₂ ACCEPTOR

FLUIDIZED BED	TEMP. (°F)	PRESSURE (psi)	REACTANTS	PRODUCT
GASIFIER	1500	150 – 300	COAL – STEAM – CaO	400 BTU/S.C.F (RAW GAS)
REGENERATOR	1900	150 – 300	CHAR – AIR – DOLOMITE	CALCINED DOLOMITE

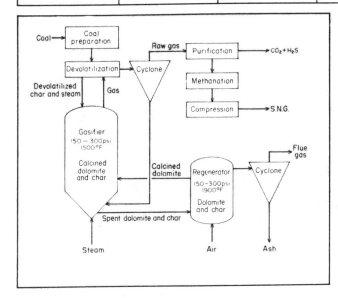

Construction of a 40 ton per day pilot plant at Rapid City, S.D. was completed in November, 1971. Testing is continuing with various feedstocks and a methanation step will be added to the pilot plant in the near future.

ELECTRIC-ARC GASIFICATION

GASIFIER	TEMP. (°F)	REACTANTS	PRODUCT GAS
ELECTRIC-ARC	18 000	COAL – STEAM	HIGH BTU S.N.G.

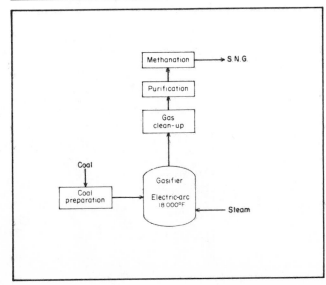

The process has been tested on a batch basis at about 30 kW. Dependent on reaction and quench conditions, the process can produce low, medium or high Btu offgas.

ELECTROFLUIDIC GASIFICATION

GASIFIER	TEMP. (°F)	PRESSURE (psi)	REACTANTS	PRODUCT GAS
FLUIDIZED BED	1500	1500	CHAR—STEAM	SYNTHESIS GAS

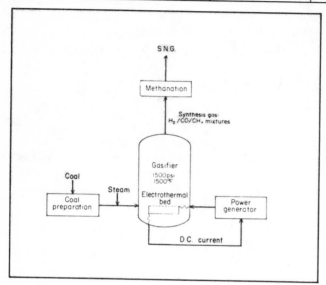

A 12 inch diameter continuous reactor has been successfully operated. The process has been used in the I.G.T. HYGAS pilot plant for generation of hydrogen-rich gas for hydrogasification.

EXXON GASIFICATION

FLUIDIZED BED	TEMP. (°F)	REACTANTS	PRODUCT GAS (RAW)
GASIFIER	1500—1700	COAL—STEAM	MED. BTU GAS FOR METHANATION
CHAR HEATER	1700+	CHAR—AIR	

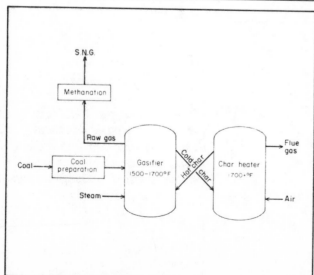

An 0.5 ton per day test unit has been in operation at Baytown, Texas for some years. Plans for a 500 ton per day pilot plant at Baytown have been deferred.

GEGAS

THE PROCESS EMPLOYS A MOVING, FIXED — BED GASIFIER

FEW DETAILS ARE AVAILABLE ON THIS PROCESS

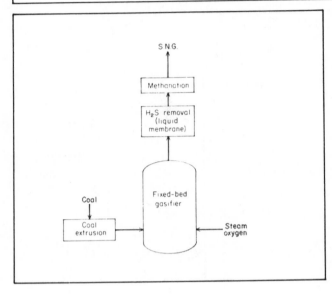

Preliminary tests have been completed in a 50 lb. per hour unit. General Electric is seeking partners for a demonstration plant.

HYDRANE

REACTOR	TEMP. (°F)	PRESSURE (psi)	REACTANTS	PRODUCT GAS
UPPER-DILUTE PHASE	1650	1000	COAL-LOWER BED GAS	MED. BTU GAS
LOWER-FLUIDIZED BED	1650	1000	CHAR-HYDROGEN	826 BTU/S.C.F. OFFGAS
H_2 GENERATOR	1800—1900	1000	CHAR-STEAM-O_2	HYDROGEN-RICH GAS

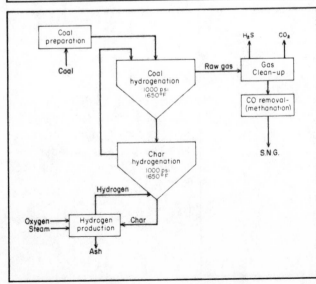

A 10 lb. per hour bench-scale unit has successfully demonstrated the feasibility of the process. Scale-up to a 24 ton per day plant is planned.

119

HYGAS

FLUIDIZED BED	TEMP. (°F)	PRESSURE (psi)	REACTANTS	PRODUCT
COAL PRETREATER	750	ATMOSPHERIC	COAL – AIR	FUEL GAS-TREATED COAL
SLURRY VAPORIZER	600	1000 – 1500	COAL – OIL	SLURRY OIL VAPORIZED
HYDROGASIFIER – STAGE 1	1300 – 1500	1000 – 1500	COAL-STAGE 2 GAS	20% OF COAL TO CH_4
HYDROGASIFIER – STAGE 2	1700 – 1800	1000 – 1500	CHAR-H_2 RICH GAS	25% OF COAL TO CH_4

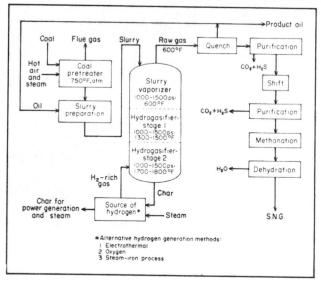

A 75 ton per day pilot plant was completed at Chicago in 1971 to produce 1·5 million S.C.F. per day of S.N.G. Testing is continuing and design data for a 250 million S.C.F. per day commercial plant will be available in June, 1975.

KOPPERS-TOTZEK

REACTOR	TEMP. (°F)	PRESSURE	REACTANTS	PRODUCT GAS (RAW)
ENTRAINED FUEL	3300	ATMOSPHERIC	COAL–STEAM–O_2	300 BTU/S.C.F.

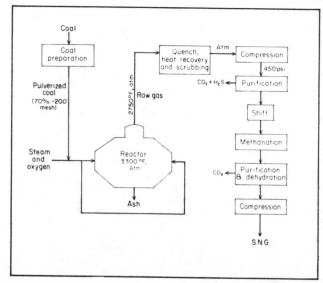

There are 16 commercial plants operating to produce a synthesis gas for ammonia production. The process is under consideration for current installation.

LIQUID-PHASE METHANATION

REACTOR	TEMP. (°F)	PRESSURE (psi)	CATALYST	CATALYST LIFE	REACTANTS	PRODUCT GAS
CATALYTIC	640	500	NICKEL	2 YEARS	$CO-H_2$	S.N.G.

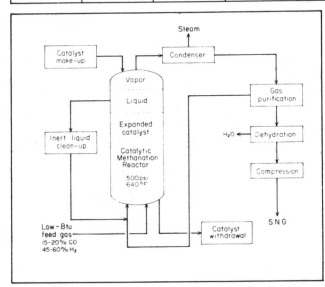

A skid-mounted pilot plant is under construction. It will have a reactor 15ft. long and 2ft. in diameter with a capacity of 2 million S.C.F. per day. The pilot plant should be ready for start-up at either the HYGAS pilot plant or the CO_2 ACCEPTOR pilot plant in May, 1975, converting synthesis gas to pipeline gas (S.N.G.).

LURGI

REACTOR	TEMP. (°F)	PRESSURE (psi)	REACTANTS	PRODUCT (OFFGAS)
FIXED BED	1140—1400	350—450	$COAL-STEAM-O_2$	450 BTU/S.C.F.

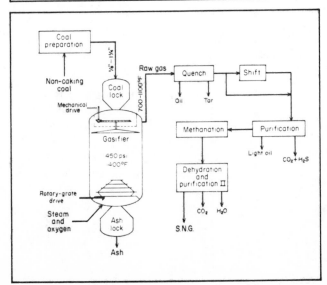

The first commercial plant was constructed in 1936. To date, 14 commercial plants have been built. About 30 Lurgi-based S.N.G. plants are planned in the U.S.A. with the first expected on stream in 1978. Expansion of the Lurgi-based SASOL plant in South Africa is planned.

MOLTEN-SALT

REACTOR	TEMP. (°F)	PRESSURE (psi)	REACTANTS	PRODUCT GAS (RAW)
CATALYTIC-MOLTEN Na_2CO_3	1830	420	COAL-STEAM-O_2	330 BTU/S.C.F.

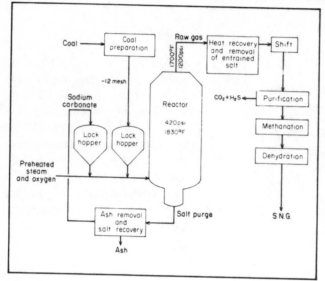

This process has been studied since 1964. Recent developments (incorporating the reaction in a single vessel without circulation of molten salt catalyst, and non-corrosive reactor lining) have increased the appeal of the process for both S.N.G. production and as a source of low-Btu fuel gas for magnetohydrodynamic generators.

MULTIPLE CATALYST

REACTOR	TEMP. (°F)	PRESSURE (psi)	REACTANTS	PRODUCT GAS
HEATED Ni CATALYST	1200—1300	1000	COAL-STEAM	850 BTU/S.C.F. AFTER PURIFICATION

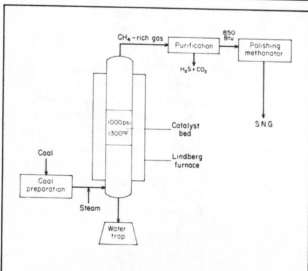

A continuous reactor has been in operation for some time. 55 different catalysts have been tested in runs from 1 to 30 hrs. in duration. Before large-scale operation is possible, a better catalyst system must be developed.

SOLUTION-GASIFICATION

```
NO DETAILS ARE AVAILABLE ON OPERATING CONDITIONS
```

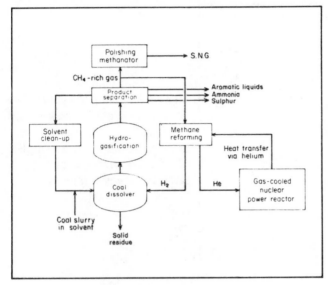

A demonstration plant is planned in Oklahoma, wherein a nuclear reactor will supply heat for the methane-reforming step which produces hydrogen for the gasifier. A conceptual design for a commercial plant processing 36 000 tons per day of coal to yield 600 million S.C.F. per day of S.N.G. has been commissioned by O.C.R.

SYNTHANE

REACTOR	TEMP. (°F)	PRESSURE (psi)	REACTANTS	PRODUCT GAS (RAW)
FLUIDIZED BED	1800	1000	COAL-STEAM-O_2	405 BTU/S.C.F.

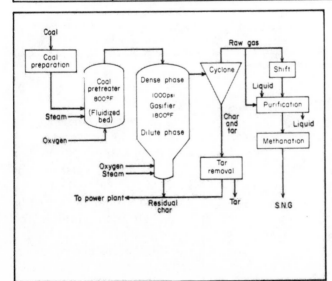

This process has been under development since 1961. A 75 ton per day pilot plant is under construction at Bruceton Pa. with completion expected in mid-fiscal year 1975. The pilot plant will produce 2·4 million S.C.F. per day of synthesis gas, to be up-graded to 1 million S.C.F. per day of S.N.G. in a methanator to be installed at the plant.

WELLMAN-GALUSHA

REACTOR	TEMP. (°F)	PRESSURE	REACTANTS	PRODUCT GAS (RAW)
REVOLVING GRATE	1200	ATMOSPHERIC	COAL−STEAM−O_2	270 BTU/S.C.F

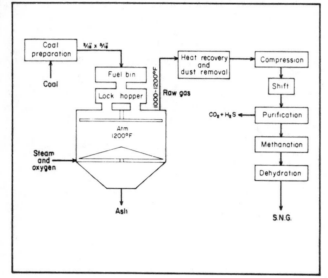

This process has been commercial for over 30 years. The vendor can only supply small units suitable for a single, large industrial plant or a complex of smaller plants requiring S.N.G. in special cases.

WINKLER

REACTOR	TEMP. (°F)	PRESSURE	REACTANTS	PRODUCT GAS (RAW)
FLUIDIZED BED	1500—1800	ATMOSPHERIC	COAL−STEAM−O_2	275 BTU/S.C.F.

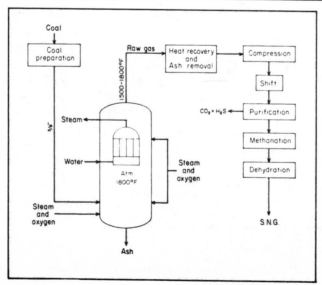

This process is commercial in 16 plants. The plants have a maximum output of 1·1 million S.C.F. per day of synthesis gas suitable for the production of ammonia, methanol and oil by Fischer-Tropsch synthesis. The last installation was in 1960 but the process is once again under consideration for current installation.

SUBJECT INDEX

Gas phase 15.
Gas removal system 24.
Gas turbines 26, 60, 61, 68, 74.
Gegas process 47, 63.
G.E.M.C.O. (see also Gulf Energy Minerals Co.) 2, 33.
Gneral Electric Co. of N.Y. 47, 60, 63.
General Electric Research & Development Co. N.Y. 47, 63.
General Tire & Rubber Co. 4, 39.
Georgia Institute of Technology 68.
Governor of Ohio 4.
Governor of West Virginia 4.
Gravity decanter vessel 6.
Groundwater seepage 66.
Great Lakes Carbon Corp. 37, 94.
Great Northern Railroad Co. 25.
Gulf Energy & Minerals Co. 2, 33.
Gulf General Atomics Co. 54.
Gulf Mineral Resources Co. 2, 18, 33.
Gulf Oil Corp. 1, 8, 18, 33, 54, 61.
Gulf Research & Development Co. 2, 94.

Hanna project 65.
Heat carrier 22.
Heat exchanger 63.
H-Coal process 13, 18, 26, 33, 90.
H-donor method 28.
Heated nickel catalyst 53.
Heating oil/fuel 14, 31.
Heavy;
    alcohol, 15.
    bottoms, 14.
    hydrogen 27.
    oil, 29.
    slurry oil, 49.
Heinrich Koppers GmbH 49, 67, 94.
Helium 68.
Hetero-atoms 31.
H.G.R. 52, 55.
High boiling point fractions 22.
High heat transfer 17.
High volatile bituminous coal 2, 7, 17, 23, 47.
H-oil process 18.
Hot char 6, 17, 22, 46.
Hot gas recycle process (H.G.R.) 52, 55.
H.R.I. fluidized bed process 63.
H.R.I. gasification process (Squires) 63.
H.R.I. Technical Services Inc. 18.
Huntington Chemical Corp. 37, 94.
Hydrane process 47.
Hydraulic fracturing 65.
    shockwave, 65.
Hydrocarbon oils 10, 18.
Hydrocarbon Research Inc. (H.R.I.) 18, 94.
Hydroclones 12, 19, 32.
Hydroconversion 31, 58.
Hydrocracking; 21, 24, 30, 31.
    catalyst, 22.
    process, 20, 65.
Hydrogen 1, 3, 6, 9, 14, 15, 17, 18, 21, 22, 27, 28, 30, 36, 37, 57.
Hydrogenated; 28.
    products, 37.
    Thianaphthenes 37.
Hydrogenation; 1, 3, 10, 14, 30, 37, 48.
    catalyst 27.
    (high pressure) 3.
    liquefaction reactor, 14.
    liquid phase, 37.
    plant, 12, 14.
    quadri phase, 37.
    residue, 3.
Hydrogasification 56.
Hydrogasification reactor 47.
Hydrogasifier 47, 56.
Hydrogen donor solvent 14, 21, 23, 26, 27, 36.
Hydrogen; 9.
    generator 4, 12, 14, 17, 47, 54, 56.
    plasma arc process, 59.
    rich gas, 30, 47.
    sulphide, 15, 30, 35, 37.
Hydrotreatment 9, 12, 17, 31.
Hydrotreating section 6, 24.
Hygas 40, 45, 48, 51, 73, 87.
    Hygas pilot plant, 48, 51.

I.C.C.R. 87.
I.C.I. moving burden process 39, 63.
I.F.E. two stage process 64.
I.G. Farben syndicate 1.
I.G.I. two stage process 64.
I.G.I. two stage (modified) 64.
Il Gaz Intergrale 64.
Illinois No. 6 Coal 6, 22, 28, 32.
Illinois Institute of Technology 48.
Imbibation 36.
Immobilized catalyst 35.
Imperial Chemical Industries Ltd. 24, 50, 63.
Indirect heat exchanger 56.
Industrial coke 25.
Industrial Development Company of Rhodesia 87.
Industrial Development Corp. (So. Africa) 16.
Inert bulk diluting agents 47.
Inert heat carrier 11.
Inert liquid phase 50.
Inerts 30, 43.
In-situ gasification 36, 43, 65, 85, 88, 90.
Insoluble residue 27.
Insoluble solids 18.
Institute of Gas Technology 45, 48, 73, 74, 87, 94.
Integrated gasifier 56.
Intermediate hydrogenation process 20.
International Furnace Equipment Co. 64.
International Oil Ltd. 90.
International System & Controls Corp. 40, 73.
Iowa Power & Light Co. 77.
Iowa State University 44.
Iron;
    Steel Industry, 41.
    oxide, 1.
    separation, 70.
Irradiation 67.
Island Creek Coal Co. 46.

Jackson & Moreland Division 76.
Jet fuel 28.

Kellogg, M.W. & Co. 8, 15, 16, 53, 62, 69.
Kellogg molten salt 67, 70, 75.
Kerosene 19.
Kerpeley producer process 67.
Kerr-McGee Corp. 94.
Koppers Company (see Heinrich Koppers GmbH)
Koppers recycling process 65.
Koppers Totzek gasifier 50.
Koppers Totzek process 7, 8, 49, 58, 67, 77.

Laser 67.
Laser irradiation pyrolysis process 67.
Latent heat 50.
Lawrence Livermore Laboratory 66.
Lawrence plant 61.
Lead time 9.
Leas Bros. Development Corp. 21, 94.
Leas liquefaction process 21.
Leuna synthesis gas process 75.
Linberg furnace 53.
Lift tube 11, 22.
Light hydrocarbons 27.
Light oil 31.
Lignite 4, 13, 21, 23, 24, 25, 41, 42, 43, 48.
Lignite refinery concept 25.
Lime sorbent 76.
Liquefaction; 1, 25.
    reaction, 24.
    reactor, 14.
    section, 9.
Liquefied Gas 1.
Liquid;
    coal extract, 27.
    fuel, 31.
    Fuels Advisory Board (S.A.) 16.
    hydrocarbons, 15.
    phase hydrogenation, 37.
    phase methanation (L.P.M.), 43, 50.
    products, 2, 12, 14, 17, 26.
    selective membranes, 47.
    solvent section, 18.
Liquids treatment section 2.
Liquor 15.